12

Algorithmic Adventures

Juraj Hromkovič

Algorithmic Adventures

From Knowledge to Magic

Prof. Dr. Juraj Hromkovič
ETH Zentrum
Department of Computer Science
Swiss Federal Institute of Technology
8092 Zürich
Switzerland
juraj.hromkovic@inf.ethz.ch

ISBN 978-3-540-85985-7 e-ISBN 978-3-540-85986-4
DOI 10.1007/978-3-540-85986-4
Springer Dordrecht Heidelberg London New York

Library of Congress Control Number: 2009926059

ACM Computing Classification (1998): K.3, A.1, K.2, F.2, G.2, G.3

Cover design: KünkelLopka GmbH, Heidelberg

Printed on acid-free paper

Springer is part of Springer Science+Business Media (www.springer.com)

For

Urs Kirchgraber

Burkhard Monien

Adam Okrúhlica

Péťa and Peter Rossmanith

Georg Schnitger

Erich Valkema

Klaus and Peter Widmayer

and all those

who are fascinated

by science

Science is an innerly compact entity.
Its division into different subject areas
is conditioned not only by the essence of the matter
but, first and foremost, by the limited
capability of human beings in the process of getting insight.

Max Planck

Preface

The public image of computer science does not reflect its true nature. The general public and especially high school students identify computer science with a computer driving license. They think that studying computer science is not a challenge, and that anybody can learn it. Computer science is not considered a scientific discipline but a collection of computer usage skills. This is a consequence of the misconception that teaching computer science is important because almost everybody possesses a computer. The software industry also behaved in a short-sighted manner, by putting too much pressure on training people in the use of their specific software products,

Searching for a way out, ETH Zurich offered a public lecture series called *The Open Class — Seven Wonders of Informatics* in the fall of 2005. The lecture notes of this first Open Class were published in German by Teubner in 2006. Ten lectures of this Open Class form the basis of *Algorithmic Adventures.*

The first and foremost goal of this lecture series was to show the beauty, depth and usefulness of the key ideas in computer science. While working on the lecture notes, we came to understand that one can recognize the true spirit of a scientific discipline only by viewing its contributions in the framework of science as a whole. We present computer science here as a fundamental science that, interacting with other scientific disciplines, changed and changes our view on the world, that contributes to our understanding of the fundamental concepts of science and that sheds new light on and brings new meaning to several of these concepts. We show that computer science is a discipline that discovers spectacular, unexpected facts, that finds ways out in seemingly unsolvable situations, and that can do true wonders. The message of this book is that computer science is a fascinating research area with a big impact on the real world, full of spectacular ideas and great challenges. It is an integral part of science and engineering with an above-average dynamic over the last 30 years and a high degree of interdisciplinarity.

The goal of this book is not typical for popular science writing, which often restricts itself to outlining the importance of a research area. Whenever possible we strive to bring full understanding of the concepts and results presented. We take the readers on a voyage where they discover fundamental computer science concepts, and we help them to walk parts of this voyage on their own and to experience the great enthusiasm of the original explorers. To achieve this, it does not suffice to provide transparent and simple explanations of complex matters. It is necessary to lure the reader from her or his passive role and to ask her or him frequently to do some work by solving appropriate exercises. In this way we deepen the understanding of the ideas presented and enable readers to solve research problems on their own.

All selected topics mark milestones in the development of computer science. They show unexpected turns in searching for solutions, spectacular facts and depth in studying the fundamental concepts of science, such as determinism, causality, nondeterminism, randomness, algorithms, information, computational complexity and automation.

I would like to express my deepest thanks to Yannick Born and Robin Künzler for carefully reading the whole manuscript and for their numerous comments and suggestions as well as for their technical support with LaTeX. I am grateful to Dennis Komm for his technical support during the work on the final version. Very special thanks go to Jela Skerlak for drawing most of the figures and to Tom Verhoeff for his comments and suggestions that were very useful for improving the presentation of some central parts of this book. The excellent cooperation with Alfred Hofmann and Ronan Nugent of Springer is gratefully acknowledged. Last but not least I would like to cordially thank Ingrid Zámečniková for her illustrations.

Zürich, February 2009 Juraj Hromkovič

Contents

By living always with enthusiasm,
being interested in everything that seems inaccessible,
one becomes greater by striving constantly upwards.
The sense of life is creativity,
and creativity itself is unlimited.

Maxim Gorky

Chapter 1

A Short Story About the Development of Computer Science or Why Computer Science Is Not a Computer Driving Licence

1.1 What Do We Discover Here?

The goal of this chapter differs from the following chapters that each are devoted to a specific technical topic. Here, we aim to tell the story of the foundation of computer science as an autonomous research discipline in an understandable and entertaining way. Trying to achieve this goal, we provide some impressions about the way in which sciences develop and what the fundamental building blocks of computer sciences look like. In this way we get to know

J. Hromkovič, *Algorithmic Adventures*, DOI 10.1007/978-3-540-85986-4_1,

some of the goals of the basic research in computer science as well as a part of its overlap with other research disciplines. We also use this chapter to introduce all ten subjects of the following chapters in the context of the development of computer science.

1.2 Does the Building of Science Sit on Unsteady Fundamentals?

To present scientific disciplines as collections of discoveries and research results gives a false impression. It is even more misleading to understand science merely in terms of its applications in everyday life. What would the description of physics look like if it was written in terms of the commercial products made possible by applying physical laws? Almost everything created by people —from buildings to household equipment and electronic devices— is based on knowledge of physical laws. However, nobody mistakes the manufacturer of TVs or of computers or even users of electronic devices for physicists. We clearly distinguish between the basic research in physics and the technical applications in electrical engineering or other technical sciences. With the exception of the computer, the use of specific devices has never been considered a science.

Why does public opinion equate the facility to use specific software packages to computer science? Why is it that in some countries teaching computer science is restricted to imparting ICT skills, i.e., to learning to work with a word processor or to search on the internet? What is the value of such education, when software systems essentially change every year? Is the relative complexity of the computer in comparison with other devices the only reason for this misunderstanding?

Surely, computers are so common that the number of car drivers is comparable to the number of computer users. But why then is driving a car not a subject in secondary schools? Soon, mobile phones will become small, powerful computers. Do we consider introducing the subject of using mobile phones into school education? We hope that this will not be the case. We do not intend to answer these rhetorical questions; the only aim in posing them is to expose the extent of public misunderstanding about computer science. Let us only note that experience has shown that teaching the use of a computer does not necessarily justify a new, special subject in school.

The question of principal interest to us is: *"What is computer science?"* Now it will be clear that it is not about how to use a computer[1]. The main problem with answering this question is that computer science itself does not provide a clear picture to people outside the discipline. One cannot classify computer science unambiguously as a metascience such as mathematics, or as a natural science or engineering discipline. The situation would be similar to merging physics and electrical and mechanical engineering into one scientific discipline. From the point of view of software development, computer science is a form of engineering, with all features of the technical sciences, covering the design and development of computer systems and products. In contrast, the fundamentals of computer science are closely related to mathematics and the natural sciences. Computer science fundamentals play a similar role in software engineering as theoretical physics plays in electrical and mechanical engineering.

It is exactly this misunderstanding of computer science fundamentals among the general public that is responsible for the bad image of computer science. Therefore we primarily devote this book to explaining some fundamental concepts and ideas of computer science.

We already argued that the best way to understand a scientific discipline is not by dealing with its applications. Moreover, we said in the beginning that it does not suffice to view a scientific discipline as sum of its research results. Hence, the next principal questions are:

"How can a scientific discipline be founded?"

---

[1] If this were the case, then almost anybody could be considered a computer scientist.

> *"What are the fundamental building blocks of a scientific discipline?"*

Each scientific discipline has its own language, that is, its own notions and terminology. Without these one cannot formulate any claims about the objects of interest. Therefore, the process of evolving notions, aimed at approximating the precise meaning of technical terms, is fundamental to science. To establish a precise meaning with a clear interpretation usually takes more effort than making historic discoveries. Let us consider a few examples. Agreeing on the definition of the notion "probability" took approximately 300 years. Mathematicians needed a few thousand years in order to fix the meaning of infinity in formal terms[2]. In physics we frequently use the term "energy". Maybe the cuckoo knows what that is, but not me. The whole history of physics can be viewed as a never-ending story of developing our understanding of this notion. Now, somebody can stop me and say: "Dear Mr. Hromkovič, this is already too much for me. I do not believe this. I know what energy is, I learnt it at school." And then I will respond: "Do you mean the Greek definition[3] of energy as a working power? Or do you mean the school definition as the facility of a physical system to perform work? Then, you have to tell me first, what power and work are." And when you start to do that, you will find that you are running in a circle, because your definition of power and work is based on the notion of energy.

We have a similar situation with respect to the notion of "life" in biology. An exact definition of this term would be an instrument for unambiguously distinguishing between living and dead matter. We miss such a definition at the level of physics and chemistry.

Dear reader, my aim is not to unnerve you in this way. It is not a tragedy that we are unable to determine the precise meaning of some fundamental terms. In science we often work with definitions that specify the corresponding notions only in an approximate way. This is everyday business for researchers. They need

[2] We will tell this story in Chapter 3.
[3] Energy means 'power with effects' in Greek.

to realize that the meaning of their results cannot reach a higher degree of precision than the accuracy of the specification of terms used. Therefore, researchers continuously strive to transform their knowledge into the definition of fundamental notions in order to get a better approximation of the intended meaning. An excellent example of progress in the evolution of notions, taken from the history of science, is the deepening of our understanding of the notion of "matter".

To understand what it takes to define terms and how hard this can be, we consider the following example. Let us take the word "chair". A chair is not an abstract scientific object. It is simply a common object and most of us know or believe we know what it is. Now, please try to define the term "chair" by a description.

> *To* ***define*** *a term means to describe it in such an accurate way that, without having ever seen a chair and following only your description, anybody could unambiguously decide whether a presented object is a chair or not. In your definition only those words are allowed whose meaning has already been fixed.*

The first optimistic idea may be to assume that one already knows what a leg of a piece of furniture is. In this case, one could start the definition with the claim that a chair has four legs. But halt. Does the chair you are sitting on have four legs? Perhaps it only has one leg, and moreover this leg is a little bit strange[4]? Let it be! My aim is not to pester you. We only want to impress upon you that creating notions is not only an important scientific activity, but also very hard work.

We have realized that creating notions is a fundamental topic of science. The foundation of computer science as a scientific discipline is also related to building a notion, namely that of "algorithm". Before we tell this story in detail, we need to know about axioms in science.

[4] In the lecture room of OpenClass, there are only chairs with one leg in the form of the symbol "L" and the chair is fixed on a vertical wall instead of on the floor.

> ***Axioms*** *are the basic components of science. They are notions, specifications, and facts, about whose validity and truth we are strongly convinced, though there does not exist any possibility of proving their correctness.*

At first glance, this may look strange, even dubious. Do we want to doubt the reliability of scientific assertions? Let us explain the whole by using an example. One such axiom is the assumption that we think in a correct way, and so our way of arguing is reliable. Can we prove that we think correctly? How? Using our arguments, which are based on our way of thinking? Impossible. Hence, nothing else remains than to trust in our way of thinking. If this axiom is not true, then the building of science will collapse. This axiom is not only a philosophical one, it can be expressed in a mathematical way. And since mathematics is the formal language of science, one cannot do anything without it.

Let us carefully explain the exact meaning of this axiom.

If

> *an event or a fact B is a consequence of another event or fact A,*

then the following must hold:

> *if A holds (if A is true),*
> *then B holds (then B is true).*

In other words,

> *untruthfulness cannot be a consequence of truth.*

In mathematics one uses the notation

$$A \Rightarrow B$$

for the fact

> *B is a consequence of A.*

We say also

> A **implies** B.

Using this notation, the axiom says: If

both $A \Rightarrow B$, *and* A *hold,*

then

B *holds.*

We call attention to the fact that it is allowed that an untruth implies a truth. The only scenario not permitted is that untruth (falsehood) is a consequence of truth. To get a deeper understanding of the meaning of this axiom, we present the following example.

Example 1.1 We consider the following two statements A and B:

A is *"It is raining"*

and

B is *"The meadow is wet"*.

We assume that our meadow is in the open air (not covered). Hence, we may claim that the statement

"If it is raining, then the meadow is wet"

i.e.,

$A \Rightarrow B$

holds.

Following our interpretation of the terms "consequence" and "implication", the meadow must be wet (i.e., B holds) when it is raining (i.e., when A holds). Let us look at this in detail.

"A holds" means *"it is raining"*.
"A does not hold" means *"it is not raining"*.
"B holds" means *"the meadow is wet"*.
"B does not hold" means *"the meadow is dry"*.

With respect to the truth of A and B, there are the following four possible situations:

S_1: It is raining and the meadow is wet.
S_2: It is raining and the meadow is dry.
S_3: It is not raining and the meadow is wet.
S_4: It is not raining and the meadow is dry.

Usually, scientists represent these four possibilities in a so-called truth table (Fig. 1.1)

	A	B
S_1	true	true
S_2	true	false
S_3	false	true
S_4	false	false

Fig. 1.1: Truth table for A and B

Mathematicians like to try to write everything as briefly as possible and unfortunately they do it even when there is a risk that the text essentially becomes less accessible for nonspecialists. They represent truth by 1 and falsehood (untruth) by 0. Using this notation, the size of the truth table in Fig. 1.1 can be reduced to the size of the table in Fig. 1.2.

	A	B
S_1	1	1
S_2	1	0
S_3	0	1
S_4	0	0

Fig. 1.2: Truth table for A and B (short version)

It is important to observe that the truth of the implication $A \Rightarrow B$ excludes the situation S_2 in the second row (A is true and B is false) only. Let us analyze this in detail.

The first row corresponds to the situation S_1, in which both A and B hold. This means it is raining and consequently the meadow is wet. Clearly, this corresponds to the validity of $A \Rightarrow B$ and so agrees with our expectation.

The second row with "A holds" and "B does not hold" corresponds to the situation when it is raining and the meadow is dry. This situation is impossible and contradicts the truthfulness of our claim $A \Rightarrow B$, because our understanding of "$A \Rightarrow B$" means that the validity of A ("it is raining") demands the validity of B ("the meadow is wet").

The third row describes the situation S_3 in which it is not raining (A is false) and the meadow is wet (B is true). This situation is possible and the fact $A \Rightarrow B$ does not exclude this situation. Despite the first fact that it is not raining, the meadow can be wet. Maybe it was raining before or somebody watered the meadow. Or in the early morning after a cold night the dew is on the grass.

The last row (both A and B do not hold) corresponds to the situation in which it is not raining and the meadow is dry. Clearly, this situation is possible and does not contradict the validity of the claim $A \Rightarrow B$ either.

We summarize our observations. If $A \Rightarrow B$ holds and A holds ("it is raining"), then B ("the meadow is wet") must hold too. If A does not hold ("it is not raining"), then the validity of $A \Rightarrow B$ does not have any consequences for B and so B may be true or false (rows 3 and 4 in the truth table). □

When $A \Rightarrow B$ is true, the only excluded situation is

"A holds and B does not hold".

If one has a truth table for two claims A and B, in which all situations with respect to the truthfulness of A and B are possible, except the situation "A holds and B does not hold", then one can say that $A \Rightarrow B$ holds. From the point of view of mathematics, the truth table in Fig. 1.3 is the formal definition of the notion of "implication".

In this way, we have the following simple rule for recognizing and for verifying the validity of an implication $A \Rightarrow B$.

A	B	$A \Rightarrow B$
true	true	possible (true)
true	false	impossible (false)
false	true	possible (true)
false	false	possible (true)

Fig. 1.3: Definition of the implication

If in all possible situations (in all situations that may occur) in which A is true (holds), B is also true (holds), then $A \Rightarrow B$ is true (holds).

Exercise 1.1 Consider the following two statements A and B. A means: "It is winter" and B means "The brown bears are sleeping". The implication $A \Rightarrow B$ means:

"If it is winter, then the brown bears are sleeping."

Assume the implication $A \Rightarrow B$ holds. Create the truth table for A and B and explain which situations are possible and which ones are impossible.

Now, we understand the meaning of the notion of implication (of consequence). Our next question is the following one:

What have the implication and correct argumentation in common? Why is the notion of implication the basis for faultless reasoning or even for formal, mathematical proofs?

We use the notion of implication for the development of so-called **direct** argumentation (direct proofs) and **indirect** argumentation (indirect proofs). These two argumentation methods form the basic instrument for faultless reasoning. In order to make our argumentations in the rest of the book transparent and accessible for everybody, we introduce these two basic proof methods in what follows.

Consider our statements A (*"It is raining"*) and B (*"The meadow is wet"*) from Example 1.1. Consider additionally a new statement C saying that *the salamanders are happy*. We assume that $A \Rightarrow B$ holds and moreover that

$B \Rightarrow C$ (*"If the meadow is wet, then the salamanders are happy"*)

holds too. What can be concluded from that? Consider the truth table in Fig. 1.4 that includes all 8 possible situations with respect to the validity of A, B, and C.

	A	B	C	$A \Rightarrow B$	$B \Rightarrow C$
S_1	true	true	true		
S_2	true	true	false		impossible
S_3	true	false	true	impossible	
S_4	true	false	false	impossible	
S_5	false	true	true		
S_6	false	true	false		impossible
S_7	false	false	true		
S_8	false	false	false		

Fig. 1.4: Truth table for $A \Rightarrow B$, $B \Rightarrow C$

Since $A \Rightarrow B$ is true, the situations S_3 and S_4 are excluded (impossible). Analogously, the truth of $B \Rightarrow C$ excludes the possibility of the occurrence of the situations S_2 and S_6. Let us view this table from the point of view of A and C only. We see that the following situations are possible:

(i) both A and C hold (S_1)

(ii) both A and C do not hold (S_8)

(iii) A does not hold and C holds (S_5, S_7).

The situations S_2 and S_4 in which A holds and C does not hold are excluded because of $A \Rightarrow B$ and $B \Rightarrow C$. In this way we obtain

$A \Rightarrow C$ (*"If it is raining, then the salamanders are happy"*)

is true.

The implication $A \Rightarrow C$ is exactly what one would expect. If it is raining, then the meadow must be wet ($A \Rightarrow B$), and if the meadow is wet then the salamanders must be happy ($B \Rightarrow C$). Hence, through the wetness of the meadow, the rain causes the happiness of the salamanders ($A \Rightarrow C$).

The argument

"If $A \Rightarrow B$ and $B \Rightarrow C$ are true,
then $A \Rightarrow C$ is true"

is called a **direct proof** (direct argument). Direct proofs may be built from arbitrarily many implications. For instance, the truth of the implications

$$A_1 \Rightarrow A_2,\ A_2 \Rightarrow A_3,\ A_3 \Rightarrow A_4,\ \ldots,\ A_{k-1} \Rightarrow A_k$$

allows us to conclude that

$$A_1 \Rightarrow A_k$$

holds. From this point of view, direct proofs are simply sequences of correct implications. In mathematics lessons, we perform many direct proofs in order to prove various statements. Unfortunately, mathematics teachers often forget to express this fact in a transparent way and therefore we present here a small example from a mathematics lesson.

Example 1.2 Consider the linear equality $3x - 8 = 4$. Our aim is to prove that

$x = 4$ is the only solution of the equality $3x - 8 = 4$.

In other words we aim to show the truth of the implication

"If $3x - 8 = 4$ *holds, then* $x = 4$*".*

Let A be the statement "The equality $3x - 8 = 4$ holds" and let Z be the statement "$x = 4$ holds". Our aim is to prove that $A \Rightarrow Z$ holds. To do it by direct proof, we build a sequence of undoubtedly correct implications starting with A and finishing with Z.

We know that an equality[5] remains valid if one adds the same number to both sides of the equality. Adding the integer 8 to both sides of the equality $3x - 8 = 4$, one obtains

[5] To be precise, the solutions of an equality do not change if one adds the same number to both sides of the equality.

$$3x - 8 + 8 = 4 + 8$$

and consequently

$$3x = 12.$$

Let B be the assertion that the equality $3x = 12$ holds. Above we showed the truthfulness of the implication $A \Rightarrow B$ ("If $3x - 8 = 4$ is true, then $3x = 12$ is true").

Thus, we have already our first implication. We also know that an equality remains valid if one divides both sides of the equality by the same nonzero number. Dividing both sides of $3x = 12$ by 3, we obtain

$$\frac{3x}{3} = \frac{12}{3}$$

and so

$$x = 4.$$

In this way we get the truthfulness of the implication $B \Rightarrow Z$ ("If the equality $3x = 12$ holds, then the equality $x = 4$ holds").

The validity of the implications $A \Rightarrow B$ and $B \Rightarrow Z$ allows us to claim that $A \Rightarrow Z$ holds. Hence, if $3x - 8 = 4$ holds, then $x = 4$. One can easily verify that $x = 4$ satisfies the equality. Thus, $x = 4$ is the only solution of the equality $3x - 8 = 4$. □

Exercise 1.2 Show by direct argumentation (through a sequence of implications) that $x = 1$ is the only solution of the equality $7x - 3 = 2x + 2$.

Exercise 1.3 Consider the truth table for the three statements A, B, and C in Fig. 1.5.

We see that only 3 situations (S_1, S_2, and S_8) are possible and all others are impossible. Which implications hold? For instance, the implication $C \Rightarrow A$ holds, because whenever in a possible situation C is true, then A is true too. The implication $B \Rightarrow C$ does not hold, because in the possible situation S_2 it happens that B holds and C does not hold. Do you see other implications that hold?

	A	B	C	
S_1	1	1	1	
S_2	1	1	0	
S_3	1	0	1	impossible
S_4	1	0	0	impossible
S_5	0	1	1	impossible
S_6	0	1	0	impossible
S_7	0	0	1	impossible
S_8	0	0	0	

Fig. 1.5: Truth table for A, B and C in Exercise 1.3

Most of us rarely have troubles understanding the concept of direct argumentation. On the other hand indirect argumentation is considered to be less easily understood. Judging whether indirect argumentation is really more complex than direct argumentation, and to what extent the problem arises because of poor didactic approaches in schools is left to the reader. Since we apply indirect argumentation for discovering some fundamental facts in Chapters 3 and 4, we take the time to explain the schema of this reasoning in what follows.

Let us continue with our example. The statement A means "It is raining", B means "The meadow is wet", and C means "The salamanders are happy". For each statement D, we denote by $\overline{D}$ the opposite of D. In this notation, $\overline{A}$ means "It is not raining", $\overline{B}$ means "The meadow is dry", and $\overline{C}$ means "The salamanders are unhappy". Assume, as before, that the implications $A \Rightarrow B$ and $B \Rightarrow C$ hold.

Now, suppose we or the biologists recognize that

"The salamanders are unhappy"

i.e., that $\overline{C}$ holds (C does not hold). What can one conclude from that?

If the salamanders are unhappy, the meadow cannot be wet, because the implication $B \Rightarrow C$ guarantees the happiness of the salamanders in a wet meadow. In this way one knows with certainty that $\overline{B}$ holds. Analogously, the truthfulness of $\overline{B}$ and of the implication $A \Rightarrow B$ implies that it is not raining, because in the opposite case the meadow would be wet. Hence, $\overline{A}$ holds. We observe in this way that the validity of

$A \Rightarrow B$, $B \Rightarrow C$, and $\overline{C}$

implies the validity of

$\overline{B}$ and $\overline{A}$.

We can observe this argumentation in the truth table in Fig. 1.6 too. The validity of $A \Rightarrow B$ excludes the situations S_3 and S_4. Since $B \Rightarrow C$ holds, the situations S_2 and S_6 are impossible.

	A	B	C	$A \Rightarrow B$	$B \Rightarrow C$	C does not hold
S_1	true	true	true			impossible
S_2	true	true	false		impossible	
S_3	true	false	true	impossible		impossible
S_4	true	false	false	impossible		
S_5	false	true	true			impossible
S_6	false	true	false		impossible	
S_7	false	false	true			impossible
S_8	false	false	false			

Fig. 1.6: Truth table for A, B and C

Since $\overline{C}$ holds (since C does not hold), the situations S_1, S_3, S_5, and S_7 are impossible. Summarizing, S_8 is the only situation that is possible. The meaning of S_8 is that none of the statements A, B, and C holds, i.e., that all of $\overline{A}$, $\overline{B}$, and $\overline{C}$ are true. Thus, starting from the validity of $A \Rightarrow B$, $B \Rightarrow C$, and $\overline{C}$, one may conclude that $\overline{B}$, and $\overline{C}$ hold.

Exercise 1.4 Consider the statements A, B, and C with the above meaning. Assume $A \Rightarrow B$, $B \Rightarrow C$ and $\overline{B}$ hold. What can you conclude from that? Depict the truth table for all 8 situations with respect to the validity of A, B, and C and determine which situations are possible when $A \Rightarrow B$, $B \Rightarrow C$ and $\overline{B}$ hold.

We observe that the validity of $A \Rightarrow B$, $B \Rightarrow C$, and C does not help to say anything about the truthfulness of A and B. If C holds, the salamanders are happy. But this does not mean that the meadow is wet (i.e., that B holds). The salamanders can also have other reasons to be happy. A wet meadow is only one of a number of possible reasons for the happiness of the salamanders.

Exercise 1.5 Depict the truth table for A, B, and C, and determine which situations are possible when $A \Rightarrow B$, $B \Rightarrow C$, and C are true.

Exercise 1.6 Consider the following statements C and D. C means "The color yellow and the color blue were mixed", and D means "The color green is created". The implication $C \Rightarrow D$ means:

> "If one mixed the color yellow with blue, then the color green is created."

Assume that $C \Rightarrow D$ holds. Depict the truth table for C and D and explain which situations are possible and which are impossible. Can you conclude from the validity of $C \Rightarrow D$ the validity of the following statement?

> "If the created color differs from green, then the color yellow was not mixed with the color blue."

Slowly but surely, we are starting to understand the schema of indirect argumentation. Applying the schema of the direct proof, we know that a statement A holds and we aim to prove the validity of a statement Z. To do this, we derive a sequence of correct implications

$$A \Rightarrow A_1,\ A_1 \Rightarrow A_2,\ \ldots,\ A_{k-1} \Rightarrow A_k,\ A_k \Rightarrow Z$$

that guarantees us the validity of $A \Rightarrow Z$. From the validity of A and of $A \Rightarrow Z$ we obtain the truthfulness of Z.

The schema of an indirect proof can be expressed as follows.

Initial situation: We know that a statement D holds.
Aim: To prove that a statement Z holds.

We start from $\overline{Z}$ as the opposite of Z and derive a sequence of correct implications

$$\overline{Z} \Rightarrow A_1,\ A_1 \Rightarrow A_2,\ \ldots,\ A_{k-1} \Rightarrow A_k,\ A_k \Rightarrow \overline{D}.$$

This sequence of implications ends with $\overline{D}$ that clearly does not hold, because we assumed that D holds.

From this we can conclude that $\overline{Z}$ does not hold, i.e. that Z as the opposite of $\overline{Z}$ holds.

The correctness of this schema can be checked by considering the truth table in Fig. 1.7.

	D	Z	$\overline{D}$	$\overline{Z}$	$\overline{Z} \Rightarrow \overline{D}$	D holds
S_1	true	true	false	false		
S_2	true	false	false	true	impossible	
S_3	false	true	true	false		impossible
S_4	false	false	true	true		impossible

Fig. 1.7: Truth table for D and Z

The situation S_2 is impossible, because $\overline{Z} \Rightarrow \overline{D}$ holds. Since D holds[6], the situations S_3 and S_4 are impossible. In the only remaining situation S_1 the statement Z is true, and so we have proved the aimed validity of Z.

This proof method is called indirect, because in the chain of implications we argue in the opposite direction (from the end to the beginning). If $\overline{D}$ does not hold (i.e., if D holds), then $\overline{Z}$ cannot hold and so Z holds.

In our example we had $D = \overline{C}$, i.e., we knew that the salamanders are not happy. We wanted to prove that the consequence is that it is not raining, i.e., our aim was to show that $Z = \overline{A}$ holds. Expressing the implications

$$A \Rightarrow B, \, B \Rightarrow C$$

in our new notation one obtains

$$\overline{Z} \Rightarrow B, \, B \Rightarrow \overline{D}.$$

From the validity of $\overline{Z} \Rightarrow \overline{D}$ and D, we were allowed to conclude that the opposite of $\overline{Z} = A$ must hold. The opposite of $\overline{Z}$ is $Z = \overline{A}$. Hence, we have proved that it is not raining (i.e., that $\overline{A}$ holds).

The general schema of the indirect proofs is as follows. One wants to prove the truth of a statement Z. We derive a chain of implications

[6] This was our starting point.

$$\overline{Z} \Rightarrow A_1,\ A_1 \Rightarrow A_2,\ \ldots,\ A_k \Rightarrow U,$$

which in turn provides

$$\overline{Z} \Rightarrow U,$$

i.e., that $\overline{Z}$ as the opposite of our aim Z implies a nonsense U. Since the nonsense U does not hold, the statement $\overline{Z}$ does not hold, too. Hence, Z as the opposite of $\overline{Z}$ holds.

Exercise 1.7 Let x^2 be an odd integer. We want to give an indirect proof that then x must be an odd integer, too. We use the general schema for indirect proofs. Let A be the statement that "x^2 is odd" and let Z be the statement that "x is odd". Our aim is to prove that Z holds, if A is true. One can prove the implication $\overline{Z} \Rightarrow \overline{A}$ by showing that for each even number $2i$

$$(2i)^2 = 2^2 i^2 = 4i^2 = 2(2i^2)$$

and so $(2i)^2$ is an even number. Complete the argumentation of this indirect proof.

Exercise 1.8 Let x^2 be an even integer. Apply the schema of indirect proofs in order to show that x is even.

Exercise 1.9 (Challenge) Prove by an indirect argument that $\sqrt{2}$ is not a rational number. Note that rational numbers are defined as numbers that can be expressed as fractions of integers.

In fact, one can view the axioms of correct argumentation as creating the notion of implication in a formal system of thinking. Usually, axioms are nothing other than fixing the meaning of some fundamental terms. Later in the book, we will introduce the definition of infinity that formalizes and fixes our intuition about the meaning of the notion of infinity. Clearly, it is not possible to prove that this definition exactly corresponds to our intuition. But there is a possibility of disproving an axiom. For instance, one finds an object, that following our intuition, has to be infinite but with respect to our definition it is not. If something like this happens, then one is required to revise the axiom.

A revision of an axiom or of a definition is not to be viewed as a disaster or even as a catastrophe. Despite the fact that the change of a basic component of science may cause an extensive reconstruction of the building of science, we view the revision as a pleasing event because the resulting new building of science is more stable and so more reliable.

Up till now, we spoke about basic components of science only. What can be said about those components that are above the base? Researchers try to build science carefully, in such a way that the correctness of the axioms (basic components) assures the correctness of the whole building of science. Especially mathematics is systematically created in this way. This is the well-known objectivity and reliability of science. At least in mathematics and sciences based on arguments of mathematics the truthfulness of axioms implies the validity of all results derived.

1.3 Origin of Computer Science as the End of Euphoria

At the end of the nineteenth century, society was in a euphoric state in view of the huge success of science, resulting in the technical revolution that transformed knowledge into the ability to develop advanced machines and equipment. The products of the creative work of scientists and engineers entered into everyday life and essentially increased the quality of life. The unimaginable became reality. The resulting enthusiasm of scientists led not only to significant optimism, but even to utopian delusions about man's capabilities. The image of the causal (deterministic) nature of the world was broadly accepted. People believed in the existence of an unambiguous chain of causes and their effects like the following one

$$\begin{aligned}\text{cause}_1 \Rightarrow \text{effect}_1 & \\ \text{effect}_1 = \text{cause}_2 & \\ \text{cause}_2 \Rightarrow \text{effect}_2 & \\ \text{effect}_2 = \text{cause}_3 & \\ \text{cause}_3 \Rightarrow \ldots &\end{aligned}$$

and attempted to explain the functioning of the world in these terms. One believed that man is able to discover all laws of nature and that this knowledge suffices for a complete understanding of the world. The consequence of this euphoria in physics was that one played the Gedankenexperiment in which one believes in the existence of so-called demons, who are able to calculate and so predict the future. Physicists were aware of the fact that the universe consists of a huge number of particles and that nobody is able to record the positions and the states of all of them at a single moment in time. Therefore, they knew that knowing all the laws of nature does not suffice for a man to predict the future. Hence, physicists introduced the so-called demons as superhumans able to record the description of the state of the whole universe (the states of all particles and all interactions between them). Knowing all the laws of nature, the hypothetical demon has to be able to calculate and so to predict the future. I do not like this idea and do not consider it to be optimistic, because it means that the future is already determined. Where then is place for our activities? Are we unable to influence anything? Are we in the best case only able to predict this unambiguously determined future? Fortunately, physics itself smashed this image. First, chaos theory showed that there exist systems such that an unmeasurably small change in their states causes a completely different development in their future. This fact is well known as the so-called butterfly effect. The final reason for ending our belief in the existence of demons is related to the discovery of the laws of quantum mechanics[7] that became a fundament of correct physics. Quantum mechanics is based on genuinely random, hence unpredictable, events that are a crucial part of the laws governing the behavior of particles. If one accepts this theory (up to now, no contradiction has ever been observed between the predictions of quantum mechanics and the experiments trying to verify them), then there does not exist any unambiguously determined future and so there is some elbow room for shaping the future.

The foundation of computer science is related to another "unrealistic" delusion. David Hilbert, one of the most famous mathe-

[7] We provide more information about this topic in Chapter 6 on randomness and in Chapter 9 on quantum computing.

maticians, believed in the existence of a method for solving all mathematical problems. More precisely, he believed

(i) that all of mathematics can be created by starting from a finite collection of suitable axioms,

(ii) that mathematics created in this way is complete in the sense that every statement expressible in the language of mathematics can be either proved or disproved in this theory,

(iii) and that there exists a method for deciding the correctness of any statement.

The notion of "method" is crucial for our consideration now. What was the understanding of this term at that time?

> *A* ***method*** *for solving a problem (a task) describes an effective path that leads to the problem solution. This description must consist of a sequence of instructions that everybody can perform (even people who are not mathematicians).*

The main point is to realize that one does not need to understand why a method works and how it was discovered in order to be able to apply it for solving given problem instances. For instance, consider the problem of solving quadratic equations of the following form:

$$x^2 + 2px + q = 0$$

If $p^2 - q \geq 0$, then the formulae

$$x_1 = -p + \sqrt{p^2 - q}$$
$$x_2 = -p - \sqrt{p^2 - q}$$

describe the calculation of the two solutions of the given quadratic equation. We see that one can compute x_1 and x_2 without any knowledge about deriving these formulae and so without understanding why this way of computing solutions to quadratic equation works. One only needs the ability to perform arithmetic operations. In this way, a computer as a machine without any intelligence can solve quadratic equations by applying this method.

Therefore, one associates the existence of a mathematical method for solving a problem to the possibility of calculating solutions in an **automatic** way. Today, we do not use the notion "method" in this context, because this term is used in many different areas with distinct meanings. Instead, we use the term **algorithm**. The choice of this new term as a synonym for a solution method was inspired by the name of the Arabic mathematician al-Khwarizmi, who wrote a book about algebraic methods in Baghdad in the ninth century. Considering this interpretation of the notion of algorithm, David Hilbert strove to automate the work of mathematicians. He strove to build a perfect theory of mathematics in which one has a method for verifying the correctness of all statements expressible in terms of this mathematics. In this theory, the main activity of mathematicians devoted to the creation of proofs would be automated. In fact, it would be sad, if creating correct argumentations—one of the hardest intellectual activities—could be performed automatically by a dumb machine.

Fortunately, in 1931, Kurt Gödel definitively destroyed all dreams of building such a perfect mathematics. He proved by mathematical arguments that a complete mathematics, as desired by Hilbert, does not exist and hence cannot be created. Without formulating these mathematical theorems rigorously, we present the most important statement in an informal way:

(a) There does not exist any complete, "reasonable" mathematical theory. In each correct and sufficiently "large" mathematical theory (such as current mathematics) one can formulate statements, whose truthfulness cannot be verified inside this theory. To prove the truthfulness of such theorems, one must add new axioms and so build a new, even larger theory.

(b) A method (algorithm) for automatically proving mathematical theorems does not exist.

If one correctly interprets the results of Gödel, one realizes that this message is a positive one. It says that building mathematics as a formal language of science is an infinite process. Inserting a new axiom means adding a new word (building a new notion) to our vocabulary. In this way, on one side the expressive power of the language of science grows, and on the other hand the power of argumentation grows, too. Due to new axioms and the related new terms, we can formulate statements about objects and events we were not able to speak about before. And we can verify the truthfulness of assertions that were not checkable in the old theory. Consequently, the verification of the truthfulness of statements cannot be automated.

The results of Gödel have changed our view on science. We understand the development of science more or less as the process of developing notions and of discovering methods. Why were the results of Gödel responsible for the founding of computer science? Here is why. Before Gödel nobody saw any reason to try and give an exact definition of the notion of a method. Such a definition was not needed, because people only presented methods for solving particular problems. The intuitive understanding of a method as an easily comprehensible description of a way of solving a problem was sufficient for this purpose. But when one wanted to prove the nonexistence of an algorithm (of a method) for solving a given problem, then one needed to know exactly (in the sense of a rigorous mathematical definition) what an algorithm is and what it is not. Proving the nonexistence of an object is impossible if the object has not been exactly specified. First, we need to know exactly what an algorithm is and then we can try to prove that, for some concrete problems, there do not exist algorithms solving them. The first formal definition of an algorithm was given by Alan Turing in 1936 and later further definitions followed. The most important fact is that all reasonable attempts to create a definition of the notion of algorithm led to the same meaning of this term in the sense of specifying the classes of automatically (algorithmically) solvable problems and automatically unsolvable problems. Despite the fact that these definitions differ in using different mathematical approaches and formalisms, and so are expressed in different ways, the class of algorithmically solvable problems determined by them is always the same. This confirmed the belief in the reasonability of these definitions and resulted in viewing Turing's definition of an algorithm as the first[8] axiom[9] of computer science.

Now, we can try to verify our understanding of axioms once again. We view the definition of an algorithm as an axiom, because it is impossible to prove its truthfulness. How could one prove that our rigorous definition of algorithmic solvability really corresponds to our intuition, which is not rigorous? We cannot exclude the possibility of a refutation of this axiom. If somebody designs a method for a special purpose and this method corresponds to our intuitive image of an algorithm, but is not an algorithm with respect to our definition, then the definition was not good enough and we have to revise it. In spite of many attempts to revise the definition of an algorithm since 1936, each attempt only confirmed Turing's definition and so the belief in this axiom grew. After proving that the class of problems algorithmically solvable by quantum computers is the same as the class of problems solvable by Turing's algorithms, almost nobody sees any possibility of a violation of this axiom.

The notion of an algorithm is so crucial for computer science that we do not try to explain its exact meaning in a casual manner now. Rather, we devote a whole chapter of this book to building the right understanding of the notions of an algorithm and of a program.

1.4 The History of Computer Science and the Concept of This Book

The first fundamental question of computer science is the following one:

[8] All axioms of mathematics are considered axioms of computer science, too.

[9] More precisely, the axiom is the claim that Turing's definition of an algorithm corresponds to our intuitive meaning of the term "algorithm".

Do there exist tasks (problems) that cannot be solved automatically (algorithmically)? And, if yes, which tasks are algorithmically solvable and which are not?

We aim not only to answer these questions, but we attempt to present the history of discovering the right answers in such a way that, following it, anybody could fully understand the correctness of these answers in detail. Since this topic is often considered to be one of the hardest subjects of the first two years of computer science study at university, we proceed towards our aim in a sequence of very small steps. Therefore, we devote to this oldest part of computer science history three whole chapters.

The second chapter is titled as follows:

"Algorithmics, or *What Do Programming and Baking Have in Common?"*

It is primarily devoted to developing and understanding the key notions of an algorithm and of a program. To get a first idea of the meaning of these terms, we start with baking a cake.

Have you ever baked a cake following a recipe? Or have you cooked a meal without any idea why the recipe asks you to work in the prescribed way? During the cooking you were aware of the fact that the correct performance of every step is enormously important for the quality of your final product. What did you discover? If you are able to follow the detailed instructions of a well-written recipe correctly, then you can be very successful in cooking without being an expert. Even if, with considerable euphoria, we may think for a moment we are masters in cooking, we are not necessarily excellent cooks. One can become a good cook only if one grasps the deeper relations between the product and the steps of its production, and can write down the recipes.

The computer has a harder life. It can perform only a few very simple activities (computation steps), in contrast to instructions present in recipes, such as mixing two ingredients or warming the content of a jar. But the main difference is that the computer does not have any intelligence and therefore is unable to improvise. A computer cannot do anything other than follow consistently step by step the instructions of its recipe, which is its program. It does not have any idea about what complex information processing it is doing.

In this way, we will discover that the art of programming is to write programs as recipes that make methods and algorithms "understandable" for computers in the sense that, by executing their programs, computers are able to solve problems. To realize this properly, we introduce a model of a computer, and show which instructions it can execute, and what happens when it is executing an instruction. In doing so, we also learn what algorithmic problems and tasks are and what the difference is between the terms algorithm and program.

The title of the third chapter is

> *"Infinity Is Not Equal to Infinity,* or *Why Is Infinity Infinitely Important for Computer Scientists?"*

This chapter is fully devoted to infinity. Why does one consider the introduction of the notion of "infinity" to be not only useful, but to be extremely important and even indispensable for understanding the functioning of our *finite* world?

The whole new universe is huge, but finite. Everything we see, everything we experiment with, and everything we can influence is finite. No one has ever been in touch with anything infinite. Nevertheless, mathematics and computer science, and therefore many other scientific disciplines are unable to exist without infinity. Already in the first class of elementary school, we meet the natural numbers $0, 1, 2, 3, 4, \ldots$ which are infinitely many.

Why does one need infinitely many numbers, when the number of all particles in the universe is a big, but still a concrete number? Why do we need larger numbers? What meaning does infinity have in computer science and how is it related to the limits of the automatically doable?

Striving to answer these questions, we will learn not only the mathematical definition of infinity, but in addition we will see why the concept of infinity is useful. We will realize that the, at first glance, artificial notion of infinity turns out to be a successful, powerful and even irreplaceable instrument for investigating our finite world.

In Chapter 4, titled

> "*Computability,* or *Why Do There Exist Tasks That Cannot Be Solved by Any Computer Controlled by Programs?*"

we first apply our knowledge about infinity to show the existence of tasks that are not algorithmically (automatically) solvable.

How can one prove the algorithmic unsolvability of concrete tasks and problems that are formulated in real life? We apply the reduction method, which is one of the most powerful and most successful tools of mathematics for problem solving. It was originally designed for getting positive results, and we use it here in a rather surprising way. We modify this method to get an instrument for producing and propagating negative results about algorithmic unsolvability of problems. In this way, we are able to present several well-motivated problems that cannot automatically be solved by means of information technology (computers). With that, the first key goal of our book—proving the existence of algorithmically unsolvable problems—is reached.

In the early 1960s, after researchers successfully developed a theory for classifying problems into automatically solvable and unsolvable ones, computers started to be widely used in industry. When applying algorithms for solving concrete problems, the question of their computational complexity and so of their efficiency became more central than the question of the existence of algorithms. Chapter 5 is devoted to the notions and concepts of complexity theory and is titled

> "*Complexity Theory,* or *What Can One Do, If the Energy of the Whole Universe Is Insufficient to Perform a Computation?*"

After the notion of an algorithm, the notion of complexity is the next key notion of computer science. 'Complexity' is understood, in computer science, as the amount of work a computer does when calculating a solution. Typically, this is measured by the number of computer operations (instructions) performed or the amount of memory used. We will also try to measure the intrinsic complexity of problems. We do so by considering the complexity of the fastest (or of the best in another sense) algorithm for solving this problem.

The main goal of complexity theory is to classify problems (algorithmic tasks) into easy and hard with respect to their computational complexity. We know that the computational complexity of problems may be arbitrarily high and so that there exist very hard problems. We know several thousand problems from practice, for which the best algorithms for solving them have to execute more operations than the number of protons in the universe. Neither the energy of the whole universe, nor the time since the Big Bang is sufficient to solve them. Does there exist a possibility to try at least something with such hard problems?

Here we outline the first miracle of computer science. There are several promising possibilities for attacking hard problems. And how to do so is the proper art of algorithmics. Many hard problems are in the following sense unstable or sensitive. A small change in the problem formulation or a small reduction in our requirements can cause a huge jump from an intractable amount of computer work to a matter of a few seconds on a common PC. How to obtain such effects is the topic of the chapters ahead.

The miracles occur when our requirements are reduced so slightly that this reduction (almost) does not matter in the applications considered, although it saves a huge amount of computer work.

The most magical effects are caused by using randomized control. The surprises are so fascinating as to be true miracles. Therefore, we devote a whole chapter to the topic of randomization:

> *"Randomness and Its Role in Nature,* or *Randomness as a Source of Efficiency in Algorithmics"*

The idea is to escape the deterministic control flow of programs and systems by allowing algorithms to toss a coin. Depending on the outcome (heads or tails), the algorithm may choose different strategies for searching for a solution. This way, one sacrifices absolute reliability in the sense of the guarantee to always compute a correct solution, because one allows some sequences of random events (coin tosses) to execute unsuccessful computations. An unsuccessful computation may be a computation without any result or even a computation with a false result. But if one can reduce the probability of executing an unsuccessful computation to one in a billion, then the algorithm may be very useful.

We call attention to the fact that in practice randomized algorithms with very small error probabilities can even be more reliable than their best deterministic counterparts. What do we mean by this? Theoretically, all deterministic programs are absolutely correct, and randomized ones may err. But the nature of the story is that the execution of deterministic programs is not absolutely reliable, because during their runs on a computer the probability of a hardware error grows proportionally with the running time of the program. Therefore a fast randomized algorithm can be more reliable than a slow deterministic one. For instance, if a randomized program computes a result in 10 seconds with an error probability 10^{-30}, then it is more reliable than a deterministic program that computes the result in 1 week. Hence, using randomization, one can obtain phenomenal gains in efficiency by accepting a very small loss in reliability. If one can jump from an intractable amount of "physical" work to a 10 second job on a common PC by paying with a merely hypothetical loss of reliability, then one is allowed to speak about a miracle. Without this kind of miracles, current Internet communication, e-commerce, and online banking would not exist.

In addition to the applications of randomness in computer science, we discuss in this chapter the fundamental questions about the existence of true randomness and we show how our attitude toward randomness has been changing in the history of science.

Chapter 7, titled

> *"Cryptography,* or *How to Transform Weak Points into Advantages"*,

tells the history of cryptography as the science of secret codes. Here, the reader finds out how cryptography became a serious scientific discipline due to the concepts of algorithmics and complexity theory. One can hardly find other areas of science in which so many miracles occur in the sense of unexpected effects, unbelievable possibilities and ways out.

Cryptography is the ancient art of creating secret codes. The goal is to encrypt texts in such a way that the resulting cryptotext can be read and understood by the legal receiver only. Classical cryptography is based on keys that are a shared secret of the sender and the receiver.

Computer science contributed essentially to the development of cryptography. First of all, it enabled us to measure the reliability (the degree of security) of designed cryptosystems. A cryptosystem is hardly breakable if every program, without knowledge of the secret key, requires an intractable amount of computer work to decrypt a cryptotext. Using this definition, we discovered cryptosystems with efficient encryption and decryption algorithms, but whose decryption is computationally hard when the secret key is unknown.

Here we will also see that the existence of hard problems not only reveals our limits, but can be useful as well. Based on this idea, so-called public-key cryptosystems were developed. They are called public-key because the key for encrypting the text into the cryptotext may be made public. The secret knowledge necessary for efficient decryption is known only to the legitimate receiver. This secret cannot be efficiently calculated from the public encryption key and so nobody else can read the cryptotext.

The next two chapters discuss the possibilities of miniaturizing computers and thereby speeding up computations, by executing computations in the world of particles and molecules.

Chapter 8 is headed

"DNA Computing, or *a Biocomputer on the Horizon"*

and is devoted to the development of biotechnologies for solving hard computing problems. Taking a simple instance of a hard problem, we show how data can be represented by DNA sequences and how to execute chemical operations on these sequences in order to "compute" a solution.

If one carefully analyzes the work of a computer, then one realizes that all computer work can be viewed as transforming texts (sequences of symbols) into other texts. Usually, the input (problem instance description) is represented by a sequence of symbols (for instance, 0's and 1's) and the output is again a sequence of symbols.

Can nature imitate such symbolic computations? DNA sequences can be viewed as sequences built from symbols `A`, `T`, `C`, and `G` and we know that DNA sequences embody information, exactly like computer data. Similar to a computer operating on its data, chemical processes can change biological data. What a computer can do, molecules can do just as easily. Moreover, they can do it a little bit faster than computers.

In Chapter 8, we explain the advantages and the drawbacks of DNA computing and discuss the possibilities of this biotechnology in algorithmics. This research area introduced already several surprises, and today nobody is able to predict anything about the dimension of the applicability of this technology in the next 10 years.

Probably no scientific discipline had such a big influence on our view of the world as physics. We associate physics with deep discoveries and pure fascination. Quantum mechanics is the jewel among jewels in physics. The importance of its discovery bears a resemblance with the discovery of fire in primeval times. Quantum physics derives its magic not only from the fact that the laws governing the behavior of particles seemingly contradict our physical experiences in the macro world. But this theory, at first disputed but now accepted, enables us to develop, at least hypothetically, a new kind of computing on the level of particles. We devote Chapter 9

> *"Quantum Computing,* or *Computing in the Wonderland of Particles"*

to this topic.

After discovering the possibility of computing with particles, the first question posed was whether the first axiom of computer science still holds. In other words, we asked whether quantum algorithms can solve problems that are unsolvable using classical algorithms. The answer is negative: quantum algorithms can solve exactly the same class of problems as classical algorithms. Hence, our first axiom of computer science became even more stable and more reliable. What then can be the advantage of quantum computers, if only hypothetically? There are concrete computing tasks of huge practical importance that can be solved efficiently using quantum algorithms, in spite of the fact that the best known deterministic and randomized classical algorithms for these tasks require intractable amounts of computer work. Therefore, quantum mechanics promises a very powerful computer technology. The only problem is that we are still unable to build sufficiently large quantum computers capable of handling realistic data sets. Reaching this goal is a stiff challenge for current physics. We do not strive here to present any details of quantum algorithmics, because that requires a nontrivial mathematical background, We only explain why building a quantum computer is a very hard task, what the basic idea of quantum computing is, and what unbelievable possibilities would be opened up by quantum mechanics in the design of secure cryptosystems.

Chapter 10, titled

> *"How to Come to Good Decisions for an Unknown Future,* or *How to Outwit a Cunning Adversary"*,

is a return to algorithmics as the kernel of computer science.

There are many situations in real life, in which one would like to know what can be expected in the near future. Unfortunately, we can very rarely look ahead and so we have to take decisions without knowing the future. Let us consider the management of a medical emergency center with mobile doctors. The aim of the center is to deploy doctors efficiently, although nobody knows when and from where the next emergency call will arrive. For instance, the control desk can try to minimize the average (or the maximum) waiting time of patients or to minimize the overall length of all driven routes.

One can develop various strategies for determining what a doctor has to do after handling a case: wait for the next case at the present location, or go back to the medical center, or take up another, strategically selected, waiting position. Another question is: Which doctor has to be assigned to the next emergency call? The principal question for these so-called online problems is whether there exists a reasonable strategy at all without any knowledge of the future.

All this can be viewed as a game between a strategy designer and a cunning adversary. After we take a decision, the aim of the adversary is to shape the future in such a way that our decision is as unfavorable as possible. Does the online strategy designer stand a chance to make a reasonable and successful decision under these circumstances? The answer varies from problem to problem. But it is fascinating to recognize that by using clever algorithmics we can often unexpectedly outwit the adversary.

1.5 Summary

Creating notions is an important and fundamental activity when founding and developing scientific disciplines. By introducing the notion of "algorithm", the meaning of the term "method" was defined exactly, and consequently computer science was founded. Thanks to the exact definition of what an algorithm is, one was able to investigate the border between the automatically (algorithmically) solvable and unsolvable, thereby demarcating the automatically doable. After successfully classifying many problems with respect to their algorithmic solvability, computational complexity became and remained up till now central to research on computer science fundamentals. The notion of computational complexity enables us to investigate the border between "practical" solvability and "impractical" solvability. It offered the basis for defining the security of cryptographic systems and so provided the fundamentals for the development of modern public-key cryptography. The concept of computational complexity provides us with a means to study the relative power of deterministic computations, nondeterministic computations, and randomized and quantum computations, and to compare them with respect to their efficiency. This way, computer science has contributed to a deeper understanding of general paradigm notions such as

> *determinism, nondeterminism, randomness, information, truth, untruth, complexity, language, proof, knowledge, communication, algorithm, simulation, etc.*

Moreover, computer science also gives a new dimension and new content to these notions, influencing their meaning. The most spectacular discoveries of computer science are mainly related to attempts at solving hard problems. This led to the discovery of many magical results in algorithmics, to which we devote the remainder of this book.

Solutions to Some Exercises

Exercise 1.3 As we see in the truth table in Fig. 1.5, only the situations S_1, S_2 and S_8 are possible. The question is, which implications hold. To answer this question, we use the following rule:

> *If the statement Y is true in all situations in which the statement X is true, then $X \Rightarrow Y$ holds. The implication $X \Rightarrow Y$ does not hold if there is a possible situation in which Y holds, but X does not hold.*

We look at $A \Rightarrow B$ first. A is true in the situations S_1 and S_2, in which B holds too. Hence, we conclude that the implication $A \Rightarrow B$ holds.
Now, let us consider the implication $B \Rightarrow A$. The statement B holds in S_1 and S_2 only, and in these situations the statement A holds, too. Hence, $B \Rightarrow A$ holds.
We consider $A \Rightarrow C$ now. A holds in S_1 and S_2. But the statement C does not hold

in the situation S_2. Hence, the implication $A \Rightarrow C$ does not hold.
In contrast, the opposite implication $C \Rightarrow A$ holds, because C holds only in the situation S_2, in which A holds too.
In this way, one can determine that the implications $A \Rightarrow B$, $B \Rightarrow A$, $C \Rightarrow A$, and $C \Rightarrow B$ hold and the implications $A \Rightarrow C$ and $B \Rightarrow C$ do not hold. The implication $A \Rightarrow C$ does not hold because in the situation S_2 the statement A holds and the statement C does not hold. Analogously, one can prove that $B \Rightarrow C$ does not hold. The implications $A \Rightarrow A$, $B \Rightarrow B$, and $C \Rightarrow C$ are always true; it does not matter which situations are possible.

Exercise 1.6 First, we draw the truth table for C and D and study in which situations the implication $C \Rightarrow D$ is true.

	C	D	$C \Rightarrow D$
S_1	holds	holds	
S_2	holds	does not hold	impossible
S_3	does not hold	holds	
S_4	does not hold	does not hold	

We see that the situations S_1, S_3, and S_4 are possible. What does it mean, to take the additional information into account that "no green color was created"? This means that D does not hold, i.e., that $\overline{D}$ holds. This fact excludes the situations S_1 and S_3. Hence, the only remaining possible situation is the situation S_4, in which $\overline{C}$ and $\overline{D}$ are true. Hence, $\overline{D} \Rightarrow \overline{C}$ holds too, and we recognize that if no green color was created ($\overline{D}$ holds), then the blue color and the yellow color were not mixed ($\overline{C}$ holds).

Exercise 1.8 We consider two statements. The statement A means "x^2 is even" and the statement B means "x is even". The truthfulness of A is a known fact. Our aim is to prove the truthfulness of B. Applying the schema of the indirect proof, we have to start with $\overline{B}$. The statement $\overline{B}$ means that "x is odd". Following the definition of odd integers, we know that x can be expressed as

$$x = 2i + 1$$

for a positive integer i. Hence, the assertion "$x = 2i + 1$" holds and we denote it as A_1. In this way, we have $\overline{B} \Rightarrow A_1$. Starting from A_1, we obtain the following statement A_2 about x^2:

$$x^2 = (2i + 1)^2 = 4i^2 + 4i + 1 = 2(2i^2 + 2i) + 1 = 2m + 1.$$

We see $x^2 = 2m + 1$ for $m = 2i^2 + 2i$ (i.e., x^2 is expressed as two times an integer plus 1) and so we obtain the statement $\overline{A}$ that x^2 is an odd integer. In this way, we have proved the following sequence of implications:

$$\overline{B} \Rightarrow A_1 \Rightarrow A_2 \Rightarrow \overline{A}$$

$$x \text{ is odd} \Rightarrow x = 2i + 1 \Rightarrow x^2 = 2m + 1 \Rightarrow \; x^2 \text{ is odd.}$$

We know that x^2 is even and so that $\overline{A}$ does not hold. Hence, following the schema of the indirect proofs, we conclude that $\overline{B}$ does not hold. Therefore, B holds and we have reached our aim.

Perfection is based upon small things,
but perfection itself is no small thing at all.

Michelangelo Buonarroti

Chapter 2

Algorithmics, or What Have Programming and Baking in Common?

2.1 What Do We Find out Here?

The aim of this chapter is not to present any magic results or real miracles. One cannot read Shakespeare or Dostoyevsky in their original languages without undertaking the strenuous path of learning English and Russian. Similarly, one cannot understand computer science and marvel about its ideas and results if one has not mastered the fundamentals of its technical language.

J. Hromkovič, *Algorithmic Adventures*, DOI 10.1007/978-3-540-85986-4_2,

As we already realized in the first chapter on computer science history, the algorithm is the central notion of computer science. We do not want to take the whole strenuous path of learning all computer science terminology. We want to show that without using formal mathematical means, one can impart an intuitive meaning of the notion of an algorithm which is precise enough to imagine what algorithms are and what they are not. We start with cooking and then we discuss to what extent a recipe can be viewed as an algorithm.

After that, we directly switch to computers and view programming as a communication language between man and machine and imagine that programs are for computers understandable representations of algorithms. At the end of the chapter, you will be able to write simple programs in a machine language on your own and will understand to a fair extent what happens in a computer during the execution of computer instructions (commands).

By the way, we also learn what an algorithmic problem (task) is and that one is required to design algorithms in such a way that an algorithm works correctly for each of the infinitely many problem instances. To work correctly means to compute the correct result in a finite time. In this way, we build a bridge to Chapter 3, in which we show how important a deep understanding of the notion of infinity is for computer science.

2.2 Algorithmic Cooking

In the first chapter, we got a rough understanding of the meaning of the notion of algorithm or method. Following it, one can say:

> *An algorithm is an easily understood description of an activity leading to our goal.*

Hence, an algorithm (a method) provides simple and unambiguous advice on how to proceed step by step in order to reach our goal. This is very similar to a cooking recipe. A recipe tells us exactly what has to be done and in which order, and, correspondingly, we perform our activity step by step.

To what extent may one view a recipe as an algorithm?

To give a simple answer to this question is not easy. But, searching for an answer, we approach a better understanding of the meaning of this crucial term.

Let us consider a recipe for an apricot flan of diameter 26 cm.

Ingredients:

3	egg whites
1	pinch of salt
6	tablespoons of hot water
100g	cane sugar
3	egg yolks
1	teaspoon of lemon peel
150g	flour
1/2	teaspoon of baking powder
400g	peeled apricots

Recipe:

1. Put greaseproof paper into a springform pan!
2. Heat the oven up to 180°C!
3. Heat up 6 tablespoons of water!
4. Mix three egg whites with the hot water and a pinch of salt, beat them until you get whipped egg white!
5. Beat 100g cane sugar and 3 egg yolks until a solid cream develops!
6. Add 1 teaspoon of lemon peel to the cream and mix them together!
7. Mix 150g flour and 1/2 teaspoon of baking powder and add it to the mixture! Then stir all contents carefully using a whisk!
8. Fill the baking tin with the cream mixture!

```
 9. Place the skinned apricots on the mixture in
    a decorative way!

10. Put the baking tin into the oven for 25-30 minutes
    until it gets a light brown color!

11. Take the flan out of the oven and let it cool!
```

The recipe is available and the only question is whether we are able to bake the flan by following it. A possible answer may be that success can depend to some extent on the experience and the knowledge of the cook.

We are ready to formulate our first requirements for algorithms.

> *An algorithm has to be such an exact description of the forthcoming activity that one can successfully perform it even in the case where one does not have any idea why the execution of the algorithm leads to the given aim. Moreover, the description (algorithm) has to be absolutely unambiguous in the sense that different interpretations of the particular instructions are excluded. It does not matter who executes the algorithm, the resulting activity and so the resulting outcome must be the same, i.e., each application of the algorithm has to reach the same result.*

Now, one can start a long discussion about which of the 11 steps (instructions) of the recipe above can be viewed as unambiguous and easily understood by everybody. For instance:

- What does it mean **"to beat until then you get whipped egg white"** (step 4)?
- What does it mean **"to stir... carefully"** (step 7)?
- What does **"decorative"** mean (step 9)?
- What does **"light brown"** mean (step 10)?

An experienced cook would say: "Alright. Everything is clear, the description is going into unnecessary detail." Somebody trying to bake her/his first cake could require even more help and may even fail to execute the whole procedure on her/his own. And this can happen in spite of the fact that our recipe is a more detailed and simpler description than the usual recipes described in cookery books. What do you think about cookery book instructions such as:

- **Quickly** put a **little bit of cooked** gelatin below the cheese and **stir them thoroughly**?

We are not allowed to accept situations in which an experienced person considers the recipe to be an algorithm and the rest of the world does not. One has to search for a way in which we can get general agreement. We already know that an algorithm is a sequence of instructions that are correctly executable by any person. This means that before defining the notion of a cooking algorithm

> *we have to agree on a list of instructions (elementary operations) such that each of these instructions can be mastered by anybody willing to cook or bake.*

For instance, such a list can contain the following instructions that are possibly correctly executable by a robot that does not have any understanding of cooking and no improvization ability.

- `Put x spoons of water into a container!`
- `Separate an egg into an egg yolk and the egg white!`
- `Heat the oven up to x°C!`
- `Bake for y minutes at x°C!`
- `Weigh x g of substance A and put it into a container!`
- `Pour x l of liquid B into a pot!`
- `Stir the content of the container using a whisk for t minutes!`
- `Mix the content of the container using a fork for t minutes!`

- `Mix the content of the two containers!`
- `Pour the mixture into the baking tin!`

Certainly, you can find many further instructions that one can consider to be simple enough in the sense that we can expect that anybody is able to execute them. In what follows we try to rewrite the recipe in such a way that only simple instructions are applied.

Let us try to rewrite step 4 of our recipe into a sequence of simple instructions.

```
4.1 Put the three egg yolks into the container G.

4.2 Put 1g of salt into G.

4.3 Put 6 tablespoons of water into the pot T.

4.4 Heat the water in T up to 60°C.

4.5 Pour the contents of T into G.
```

Now, we get trouble. We do not know how to execute the instruction **"mix until the content of G becomes whipped egg white"**. A solution may be to use some experimental values. Maybe it takes 2 minutes until the mass is stiff enough. Hence, one could write:

```
4.6 Mix the content of G for 2 minutes.
```

An instruction of this kind may also be risky. The time of mixing depends on the speed of mixing, and that may vary from person to person. Hence, we would prefer to stop mixing approximately at the moment when the mass became stiff. What do we need for that? We need the ability to execute tests in order to recognize the moment at which the whipped egg white is ready. Depending on the result of the tests, we have to make a decision on how to continue. If the mass is not stiff, we have to continue to mix for a time. If the mass is stiff then the execution of step 4 is over and we have to start to execute step 5.

How can one write this as a sequence of instructions?

```
4.6 Mix the content of G for 10 seconds.
4.7 Test whether the content of G is stiff or not.
    If the answer is "YES", then continue with step 5.
    If the answer is "NO", then continue with step 4.6.
```

In this way, one returns to step 4.6 until the required state of the mass is reached. In computer science terminology, one calls steps 4.6 and 4.7 a cycle that is executed until the condition formulated in 4.7 is satisfied. To make it transparent one uses a graphic representation such as in Fig. 2.1; this is called a **flowchart**.

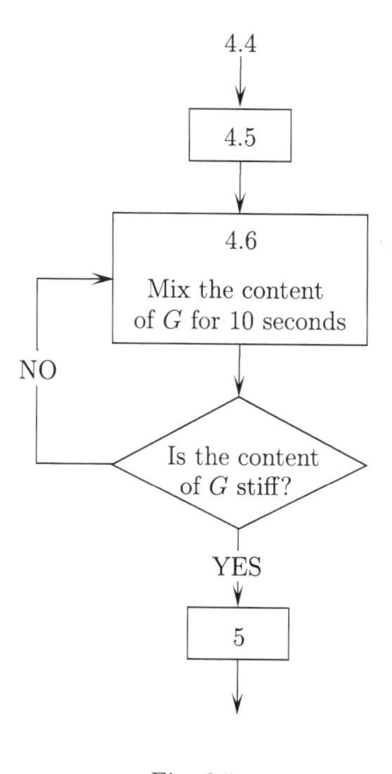

Fig. 2.1

Can one easily execute the test in step 4.6? Exactly as in the case of instructions, we have to agree on a list of simply executable tests. We do not want to discuss a possible execution of the test in step 4.6 in detail because the author is not an expert in cooking. Originally, I considered putting a teaspoon into the mass and checking whether it stays up. But a female student explained to me that this does not work and that instead it is sufficient to make a cut in the mass using a knife, and when the cut does not close (i.e., remains open) then the mass is stiff. Examples of other tests are:

- `Test whether the temperature of the liquid in a pot is at least x degrees.`
- `Test whether the weight of the content of a container is exactly x g.`

Exercise 2.1 Create a list of instructions and tests you consider anybody could execute. Then take your favorite recipe and rewrite it using the instructions and the tests from your list only.

Exercise 2.2 You want to heat 1 l of water up to 90°C. You are allowed to use only the following instructions:

- `Put the pot T on the hot plate for x seconds and then take it away.`
- `Pour x l of water into pot T.`

Moreover, you are allowed to use the following tests.

- `Test whether the water in pot T has reached at least x°C.`

Use this test and the two instructions above to write a cooking algorithm for heating 1 l of water up to 90°C that enures that the pot is not longer than 15 s on the hot plate after the water has reached 90°C.

Whether you believe it or not, after successfully solving these two exercises you have already been working as a programmer. The most important fact we learnt by baking is that we cannot speak about algorithms before the fundamental elements algorithms consist of are fixed. These elements are simple instructions and tests that everyone can execute without any problem.

2.3 What About Computer Algorithms?

Here, we want to discuss the similarities and the differences between algorithmic cooking and algorithmic computing in order to realize exactly what computer algorithms and computer programs are and what they are not.

Analogous to cooking, one has to fix first a list of fundamental instructions (operations) that a computer can execute without any doubt. To get agreement here is essentially simpler than getting it by cooking. A computer does not have any intelligence and so any improvization ability. Due to this, the language of the computer is very simple. Nobody doubts that a computer can add or multiply two integers or execute other arithmetic operations with numbers. Similarly, everyone accepts that a computer can compare two numbers as a test. These simple instructions and tests together with the ability to read the input data and to output the results are sufficient for describing any algorithm as a sequence of instructions.

It does not matter whether we consider cooking algorithms or computer algorithms. Both are nothing other than a sequence of simple instructions. But there is also an essential difference between cooking algorithms and algorithms in mathematics and in computer science. The input of a cooking algorithm is a set of ingredients and the result is a meal. The only task is to cook the aimed product from the given ingredients. Algorithmic tasks are essentially different. We know that a problem may have *infinitely many* **problem instances** as possible inputs. Consider, for instance, the problem of solving a quadratic equation.

$$ax^2 + bx + c = 0.$$

The input data are the numbers a, b, and c and the task is to find all x that satisfy this equation.

For instance, a concrete problem instance is to solve the following equation:

$$x^2 - 5x + 6 = 0.$$

Here, we have $a = 1, b = -5$, and $c = 6$. The solutions are $x_1 = 2$ and $x_2 = 3$. By substituting these values, one can easily verify that

$$2^2 - 5 \cdot 2 + 6 = 4 - 10 + 6 = 0$$
$$3^2 - 5 \cdot 3 + 6 = 9 - 15 + 6 = 0$$

and so verify that x_1 and x_2 are really the solutions of the quadratic equation $x^2 - 5x + 6 = 0$.

Because there are infinitely many numbers, one has infinitely many possibilities to choose the coefficients a, b, and c of the quadratic equation. Our clear requirements for an algorithm for solving quadratic equations is that the algorithm determines the correct solution for all possible input data a, b, and c, i.e., for each quadratic equation.

In this way, we get the second basic demand on the definition of the notion of an **algorithm**.

> *An algorithm for solving a problem (a task) has to ensure that it works correctly for each possible problem instance. To work correctly means that, for any input, it finishes its work in a finite time and produces the correct result.*

Let us consider an algorithm for solving quadratic equations. Mathematicians provided the following formulas for computing the solutions

$$x_1 = \frac{-b + \sqrt{b^2 - 4ac}}{2a}$$
$$x_2 = \frac{-b - \sqrt{b^2 - 4ac}}{2a},$$

if $b^2 - 4ac \geq 0$. If $b^2 - 4ac < 0$, there does not exist any real solution[1] to the equation. These formulas directly provide the following general method for solving quadratic equations.

[1] The reason for that is that one cannot take the root of a negative number.

Input: Numbers a, b, and c representing the quadratic equation $ax^2 + bx + c = 0$.

Step 1: Compute the value $b^2 - 4ac$.

Step 2: If $b^2 - 4ac \geq 0$, then compute

$$x_1 = \frac{-b + \sqrt{b^2 - 4ac}}{2a}$$
$$x_2 = \frac{-b - \sqrt{b^2 - 4ac}}{2a}$$

Step 3: If $b^2 - 4ac < 0$, write "there is no real solution".

Now, we believe the mathematicians when they say that this method really works and we do not need to know why in order to rewrite it as an algorithm.

However, we want to do more than to transform the description of this method into a program. The notion of a **program** is considered here as a *sequence of computer instructions* that is represented in a form that is understandable for a computer. There are essential differences between the notion of a program and the notion of an algorithm.

1. A program does not need to be a representation of an algorithm. A program may be a meaningless sequence of computer instructions.

2. An algorithm does not necessarily need to be written in the form of a program. An algorithm can also be described in a natural language or in the language of mathematics. For instance, the use of instructions such as "`multiply a and c`" or "`compute` $\sqrt{c}$" is acceptable for the description of an algorithm while a program must be expressed in a special formalism of the given programming language.

We view **programming** as an *activity of rewriting algorithms* (methods, descriptions) into programs. In what follows, we will program a little bit in order to see how one can create a complex behavior by writing a sequence of very simple instructions.

In order to be able to read and understand the forthcoming chapters, it is not necessary to study the rest of this chapter in detail. Hence, anybody not strongly interested in learning what programming is about and what happens in a computer during the execution of concrete instructions can jump this part.

We start by listing the simple operations and their representation in our programming language that we call "TRANSPARENT". In passing we show the high-level structure of a computer and see the main computer actions performed during the execution of some particular instructions.

We consider a rough, idealized model of a computer as depicted in Fig. 2.2.

This computer model consists of the following parts:

- A **memory** that consists of a large number of memory cells. These memory cells are called **registers**. The registers are numbered by positive integers and we call them **addresses** of the registers. For instance 112 is the address of `Register(112)`. This corresponds to the image in which the registers are houses on one side of a long street. Each register can save an arbitrarily large number[2].
- A special memory in which the whole program is saved. Each row of the program consists of exactly one instruction of the program. The rows are numbered starting at 1.
- There is a special register `Register(0)` that contains the number of the just executed instruction (row) of the program.
- A CPU (central processing unit) that is connected to all other parts of the computer. In order to execute one instruction,

[2] In real computers, the registers consist of a fixed number of bits, 16 or 32. The large integers or real numbers with many positions after the decimal point that cannot be represented by 32 bits have to be handled in a special way by using several registers for saving one number. Hence, we have idealized the computer here in order to remain transparent and we assume that any register can save an arbitrarily large number.

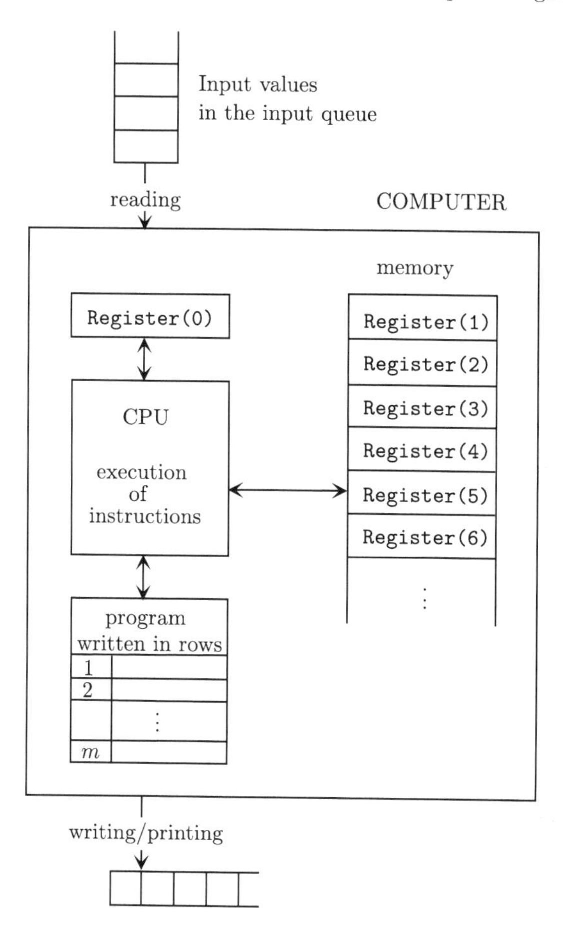

Fig. 2.2

the CPU starts by reading the content of `Register(0)` in order to fix which instruction has to be executed. Then, looking at the corresponding instruction of the program, the CPU fetches the contents of the registers (the numbers saved in the registers) that are arguments of the executed instructions and executes the corresponding operation on these data. Finally, the CPU saves the result in the register determined by the instruction and adjusts the contents of the register `Register(0)` to the number of the instruction to be executed next.

Additionally, the computer is connected to the world outside. The input data are waiting in a queue and the computer can read the first number in the queue and save it in a register. The computer has also a tape, where it can write the results computed.

Consider an analogy to baking or cooking. The computer is the kitchen. The registers of the memory are containers, bowls, jars, etc. Each container has an unambiguous name (exactly as each register has an address), and because of this one always knows which container is being considered. The memory containing the program is a sheet or a cookery book. The CPU is a person or a cookery robot together with all the other machines such as ovens, mixers, microwaves, etc. that are available in the kitchen. The content of `Register(0)` is the note telling us where we are in this process of executing the recipe. The inputs are waiting in the refrigerator or in the pantry. We have to note here that they are not waiting in a queue, but one can take all the ingredients out and build a queue that respects the order in which they are needed. Certainly we do not write the output, instead we put it on the table.

As we have already learnt by baking, the first step and the crucial point for defining the notion of an algorithm is to agree on a list of **executable** instructions (operations). Everybody has to be convinced about their executability.

In what follows, we prefer to present the possible computer instructions in natural language instead of using the formal language of the computer called machine code. We start with the instructions for reading.

(1) `Read into Register(`n`).`

To execute this operation means to take the first number of the queue and save it in `Register(`n`)`. In this way this number is deleted from the queue and the second number of the queue takes over the first position of the queue.

Example 2.1 Consider the situation in which the three numbers 114, −67, and 1 are waiting to be picked up. All registers of the memory contain the value 0, except for `Register(0)` which contains 3. One has to execute the instruction

```
Read into Register(3)
```

in the third row of the program. After the execution of this instruction, `Register(3)` contains the number 114. Then, numbers −67 and 1 are still waiting in the queue. The content of `Register(0)` is increased by 1 and becomes 4, because after executing the instruction of the third row of the program one has to continue by executing the instruction of the next row.

The execution of this instruction is depicted in Fig. 2.3. We omit describing the whole computer state before and after executing this instruction and focus on the content of the registers only. □

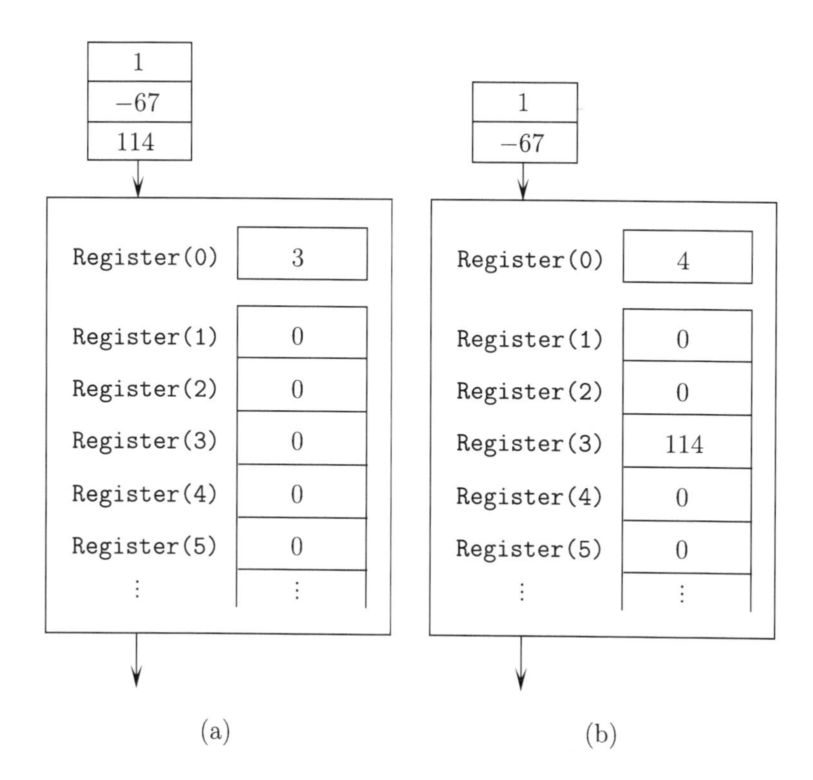

Fig. 2.3

The next instruction enables us to put a concrete number into a register without being forced to read it from the input queue

(2) `Register(`n`)` $\leftarrow k$

This instruction corresponds to the requirement to put the number k into the register `Register(`n`)`. Executing it means deleting the old content of `Register(`n`)`. After the execution of this instruction, this old content is not available anymore, it is definitely destroyed. There is no change in the queue related to this instruction.

Example 2.2 Consider that `Register(50)` contains the number 100. After executing the instruction

`Register(50)` $\leftarrow 22$

`Register(50)` contains the number 22. The old content 100 of `Register(50)` is not saved anywhere and so it is definitely lost.

If the next instruction is

`Read into Register(50)`

and the number 7 is waiting in the queue, then after the execution of this instruction the number 22 in `Register(50)` is exchanged for the number 7. □

Exercise 2.3 The numbers 11, 12, and 13 are waiting in the input queue. The content of `Register(0)` is 1. `Register(2)` contains 1117 and `Register(3)` contains 21. All other registers contain 0.

a) Depict this situation analogously to Fig. 2.3.
b) Execute the following program

```
1 Read into Register(1)
2 Register(2) ← 100
3 Read into Register(3)
4 Read into Register(2)
```

Determine and depict the content of all registers and the input queue after the execution of each particular instruction of the program.

Now we introduce some of the possible arithmetic instructions.

(3) Register(n) $\leftarrow$ Register(j) + Register(i)

The meaning of this instruction is as follows. One has to add the content of Register(j) to the content of Register(i) and to save the result in Register(n). Executing this instruction, the original content of Register(n) is overwritten by the result of the addition. The contents of all other registers remain unchanged, except for the content of Register(0) that is increased by 1 (i.e., the execution of the program has to continue with the next instruction). There is also no change in the input queue.

Example 2.3 Consider the situation (the state of the computer), in which Register(0) contains 5 and each Register(i) contains the number i for $i = 1, 2, 3, 4$, and 5 (Fig. 2.4a). All other registers contain 0. The 5th row of the program contains the following instruction:

Register(7) $\leftarrow$ Register(1) + Register(4).

Figure 2.4b shows the situation reached after the execution of this instruction of addition.

The value 1 from Register(1) and the value 4 from Register(4) are summed ($1 + 4 = 5$) and the result 5 is saved in Register(7). The contents of Register(1) and Register(4) do not change during the execution of this instruction.

Assume that row 6 of the program contains the following instruction:

Register(7) $\leftarrow$ Register(1) + Register(7).

The content of Register(1) is 1 and the content of Register(7) is 5. Accordingly, the computer computes $1 + 5 = 6$ and saves 6 in Register(7). In this way, the original content of Register(7) is deleted. Executing this instruction, we observe that one is also allowed to save the result of a computer operation in one of the two registers containing the operands (the incoming values for the operation). $\square$

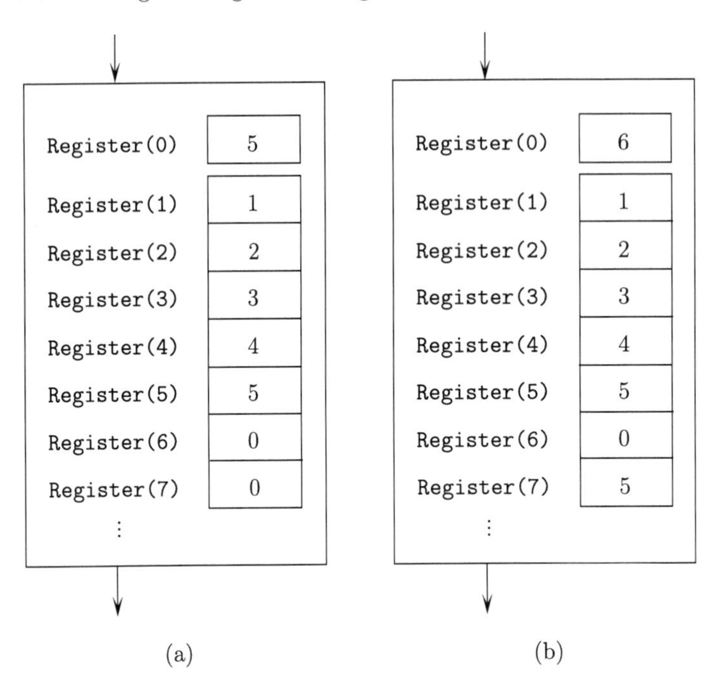

Fig. 2.4

Exercise 2.4 Consider the computer state after executing the first addition (the instruction in row 5) in Example 2.3. This situation is depicted in Fig. 2.4b. Depict the memory state (analogously to Fig. 2.4b) after executing the second addition operation from row 6! After that perform the following three instructions

```
7 Register(3) ← 101
8 Register(3) ← Register(3) + Register(3)
9 Register(3) ← Register(7) + Register(3)
```

of the program and depict the final state of the memory.

Analogously to addition one can perform other arithmetic operations too.

(4) `Register(`n`)` ← `Register(`j`) - Register(`i`)`

To execute this operation means to subtract the content of `Register(`i`)` from the content of `Register(`j`)` and to save the result in `Register(`n`)`.

(5) `Register(`n`)` ← `Register(`j`)` $*$ `Register(`i`)`

The computer has to multiply the contents of the registers `Register(`j`)` and `Register(`i`)` and to save the result in `Re-gister(`n`)`.

(6) `Register(`n`)` $\leftarrow$ `Register(`j`) / Register(`i`)`

The computer has to divide the content of `Register(`j`)` by the content of `Register(`i`)` and to save the result in `Register(`n`)`.

(7) `Register(`n`)` $\leftarrow$ $\sqrt{\texttt{Register}(m)}$

The computer has to compute[3] the root of the content of `Register(`m`)` and to save the result in `Register(`n`)`.

Exercise 2.5 Consider the following situation. All registers except[4] for `Register(0)` contain the value 0. `Register(0)` contains the value 1. The numbers a and b are waiting in the input queue. Explain what result is in `Register(3)` after the execution of the following program:

```
1 Read into Register(1)
2 Register(1) ← Register(1) * Register(1)
3 Read into Register(2)
4 Register(2) ← Register(2) * Register(2)
5 Register(3) ← Register(1) + Register(2)
```

Similarly to cooking, it is not sufficient to be able to execute some instructions only. We also need tests that decide about how to continue in the work. For this purpose, we present the following two simple basic operations:

(8) `If Register(`n`) = 0, then go to row `j

One has to test the content of `Register(`n`)`. If it is 0, the content of `Register(0)` is overwritten by the value j. This means that the execution of the program is going to continue by executing the instruction of row j. If the content of `Register(`n`)` is different from 0, then the computer adds 1 to the content of

[3] To compute a root of a number is not a basic instruction of a computer and we introduce it only because we need it for solving quadratic equations. On the other hand, there is no doubt that a computer can compute a root of a number, but to do so one has to write a program as a sequence of arithmetic instructions.

[4] Remember that `Register(0)` contains the order of the instruction executed.

`Register(0)` and the work is going to continue by executing the instruction in the next row.

(9) `If Register(`n`)` $\leq$ `Register(`m`), then go to row` j

If the content of `Register(`n`)` is not larger than the content of `Register(`m`)`, then the next instruction to be executed is the instruction of row j. Else the computer is going to execute the instruction of the next row.

The instruction (operation)

(10) `Go to row` j

is an ultimatum to continue the execution of the program in row j.

Moreover, we still need operations for outputting (displaying) the results of the computation.

(11) `Output` $\leftarrow$ `Register(`j`)`

The content of `Register(`j`)` is written (displayed) as the output.

(12) `Output` $\leftarrow$ "`Text`"

The given text between " " will be displayed. For instance, the following instruction

`Output` $\leftarrow$ "`Hallo`",

results in the word "Hallo" being written on the output tape.

The last instruction is

(13) `End`.

This instruction causes the end of the work of the computer on the given program.

Now we are ready to rewrite our algorithm for solving quadratic equations to a program. To make it transparent, we notice the current state of registers in parentheses.

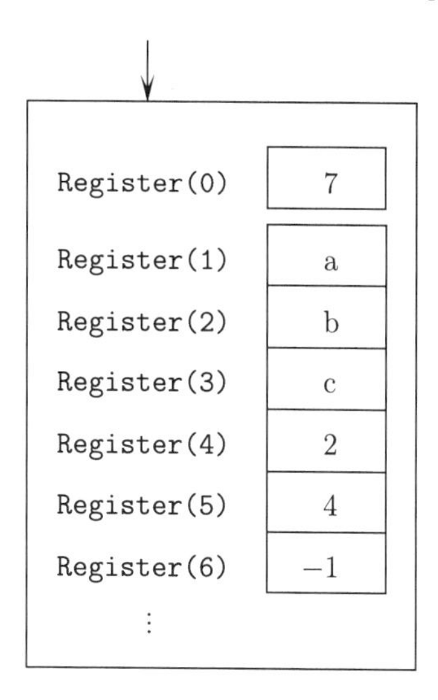

Fig. 2.5

Input: Integers a, b, c
Program:

1 `Read into Register(1)`
{`Register(1)` contains a}

2 `Read into Register(2)`
{`Register(2)` contains b}

3 `Read into Register(3)`
{`Register(3)` contains c}

4 `Register(4)` ← `2`

5 `Register(5)` ← `4`

6 `Register(6)` ← `-1`
{The state of the memory is described in Fig. 2.5}

7 `Register(7)` ← `Register(2) * Register(2)`
{`Register(7)` contains b^2}

8 `Register(8)` ← `Register(5) * Register(1)`
{`Register(8)` contains $4a$}

9 `Register(8)` $\leftarrow$ `Register(8) * Register(3)`
{`Register(8)` contains $4ac$}

10 `Register(8)` $\leftarrow$ `Register(7) - Register(8)`
{`Register(8)` contains $b^2 - 4ac$ and so the first step of the method for solving quadratic equations is finished.}

11 `If Register(9)` $\leq$ `Register(8), then go to row 14`
{Since all registers unused up to now contain the value 0, the execution of the program continues in row 14 if $b^2 - 4ac \geq 0$, i.e., if the quadratic equation has a real solution. If $b^2 - 4ac < 0$, the computation continues in the next row.}

12 `Output` $\leftarrow$ "`There is no solution.`"

13 `End`
{After displaying "There is no solution", the computer finishes the execution of the program.}

14 `Register(8)` $\leftarrow \sqrt{\texttt{Register(8)}}$
{`Register(8)` contains the value $\sqrt{b^2 - 4ac}$.}

15 `Register(7)` $\leftarrow$ `Register(2) * Register(6)`
{`Register(7)` contains the value $-b$. The original content b^2 of `Register(7)` is deleted in this way.}

16 `Register(6)` $\leftarrow$ `Register(1) * Register(4)`
{The situation is depicted in Fig. 2.6.}

17 `Register(11)` $\leftarrow$ `Register(7) + Register(8)`

18 `Register(11)` $\leftarrow$ `Register(11) / Register(6)`
{`Register(11)` contains the first solution $x_1 = \frac{-b+\sqrt{b^2-4ac}}{2a}$.}

19 `Output` $\leftarrow$ `Register(11)`

20 `Register(12)` $\leftarrow$ `Register(7) - Register(8)`

21 `Register(12)` $\leftarrow$ `Register(12) / Register(6)`
{`Register(12)` contains the second solution $x_2 = \frac{-b-\sqrt{b^2-4ac}}{2a}$.}

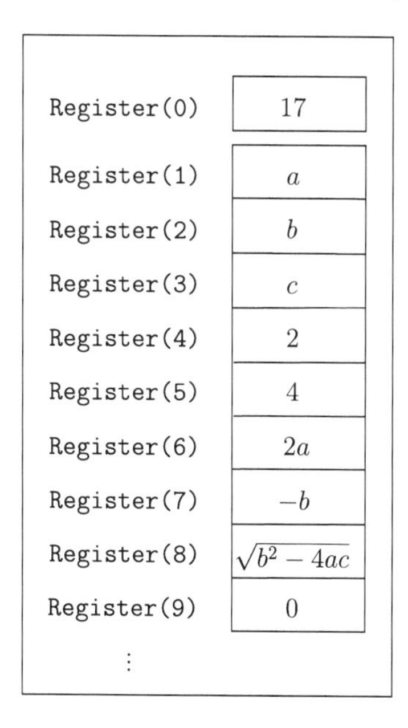

Fig. 2.6

```
22 Output ← Register(12)
23 End.
```

A transparent presentation of this program is given in Fig. 2.7.

Exercise 2.6 Describe the content of all registers after the execution of the whole program!

Exercise 2.7 If $b^2 - 4ac = 0$, then there is only one solution $x_1 = x_2$ to this quadratic equation. Modify the presented program in such a way that in this case the program outputs first the text "There is only one solution and this is" and then the value of x_1. Additionally, in the case $b^2 - 4ac > 0$ the program has to write the text "There are two solutions" before displaying x_1 and x_2.

Exercise 2.8 Explain what the following program computes!

```
1  Read into Register(1)
2  Read into Register(2)
3  Read into Register(3)
4  Register(4) ← Register(1) + Register(2)
5  Register(4) ← Register(3) + Register(4)
6  Register(5) ← 3
7  Register(6) ← Register(4) / Register(5)
```

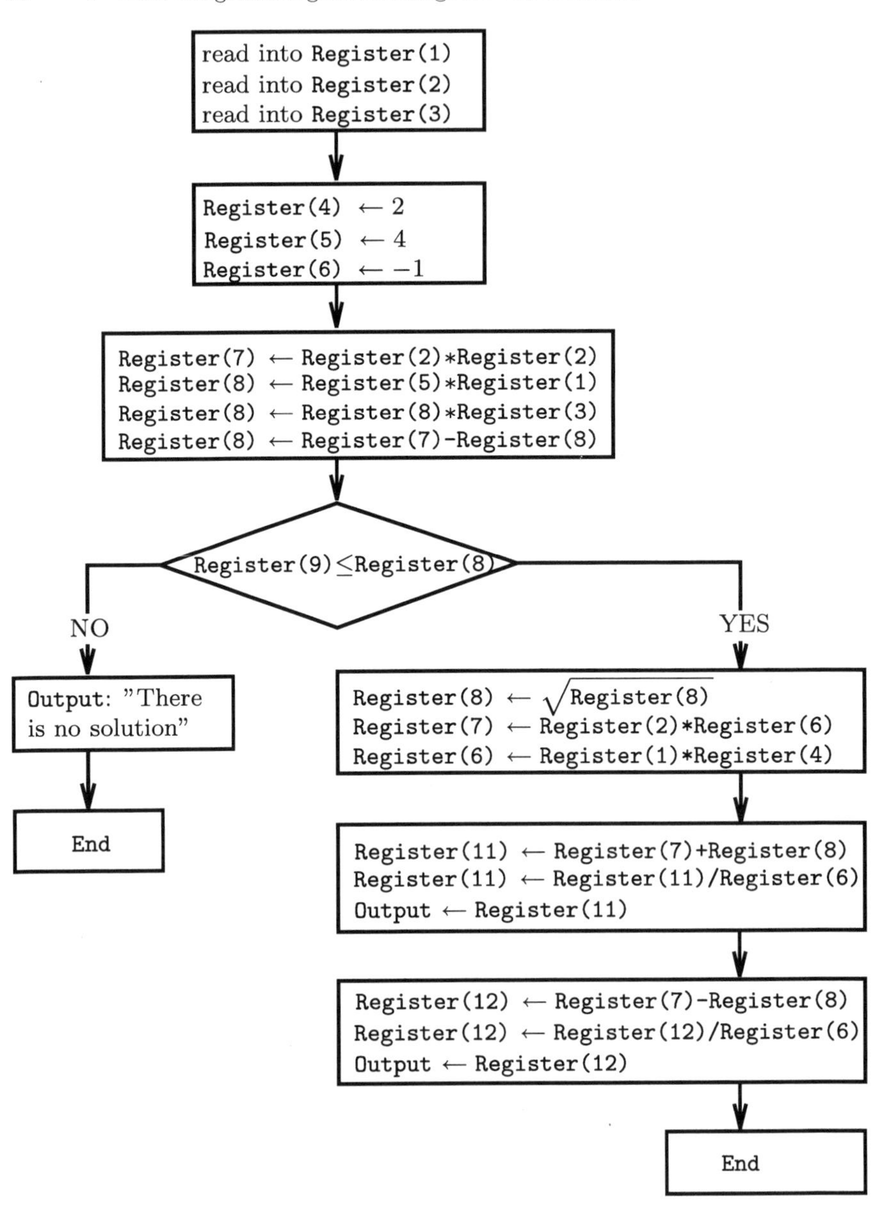

Fig. 2.7

```
8  Output ← Register(6)
9  End
```

Use a table to transparently depict the contents of the registers after the execution of the particular instructions of the program.

Exercise 2.9 Write a program that, for a given integer x, computes the following value

$$3x^2 - 7x + 11 .$$

Exercise 2.10 Write a program that, for four given numbers a, b, c, and x, computes the value

$$ax^2 + bx + c .$$

Exercise 2.11 Write a program that, for any four given integers a, b, c, and d determines and outputs the maximum of these four values.

It is always necessary to **implement** a method into a program to be able to see that the method is really an algorithm. For instance, if we see that the arithmetic operations and number comparisons are sufficient for performing the method for solving quadratic equations, then we are allowed to call this method an algorithm for solving quadratic equations. *Programming as rewriting of a method into a program is considered as a translation of an algorithm into the computer language.* From the formal point of view this transformation can be viewed as a proof of the automatic executability of the algorithm described in a natural language.

2.4 How Can the Execution of a Program Unintentionally Become a Never-Ending Story?

One of our most important demands on the definition of an algorithm for a computing task is that the algorithm finishes its work for any input and provides a result. In the formal language of computer science, we speak about **halting**. If an algorithm A finishes its work on an input (a problem instance) in a finite time, then we say that the **algorithm** $\boldsymbol{A}$ **halts on** $\boldsymbol{x}$. In this terminology, we force a halt of the algorithm on every possible input and in such a case we say that $\boldsymbol{A}$ **always halts**.

One could say now: "This is obvious. Who can be interested in developing programs for solving problems that work infinitely long and do not provide any output?". The problem is only that the algorithm designer or a programmer can unintentionally build a program that gets into an infinite repetition of a loop. How can one expect such a mistake from a professional? Unfortunately, this can happen very easily. The programmer can forget about a special situation that can appear only under very special circumstances. Let us return to our cooking algorithms to see how easily a mistake leading to an infinite work can happen.

We want to make tea by heating water and then putting the tea into the water. Our aim is to save energy and so to avoid letting the water boil for more than 20 seconds. Starting with these requirements, one can propose the cooking algorithm presented in Fig. 2.8.

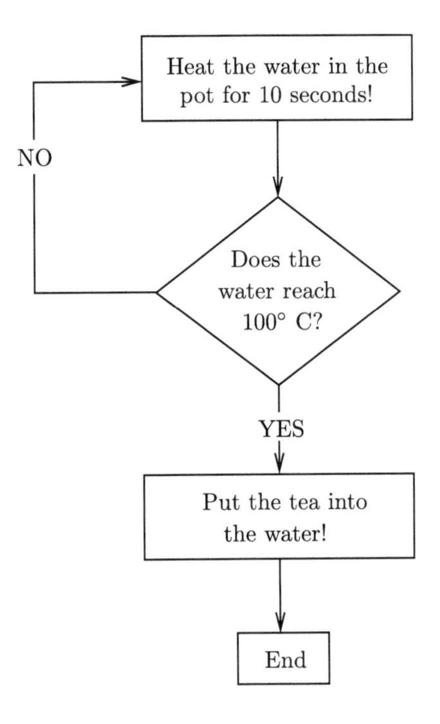

Fig. 2.8

At first glance, everything looks alright and works until a climber wants to apply this cooking recipe for preparing tea on the top of the Matterhorn on some afternoon. Do you already see the problem? Water boils at a lower temperature than 100°C at this altitude and so it may happen that it never reaches 100°C. Thus, the answer of our test will always be "NO". In reality the water won't boil forever, because eventually the fuel will run out or the water will completely vaporize.

We already see where the mistake happened. Writing the recipe, one forgot to think about this special situation, where the atmospheric pressure is so low that the water cannot reach 100°C. And the same can happen to anybody, if one does not think about all possible special problem instances of the given task and about all special situations that can occur during the computation. The following example shows such a case for a computer program.

Example 2.4 Assume, before starting the program, `Register(0)` contains 1 and all other registers contain 0. The integers a and b are waiting in the first input queue. We consider the following program.

```
1 Read into Register(1)

2 Read into Register(2)

3 Register(3) ← -1

4 If Register(1) = 0, then go to row 8

5 Register(1) ← Register(1) + Register(3)

6 Register(4) ← Register(4) + Register(2)

7 Go to row 4

8 Output ← Register(4)

9 End
```

The corresponding graphic representation of this program is presented in Fig. 2.9.

The goal of this program is to compute $a * b$. The strategy is to compute $a * b$ in the following way

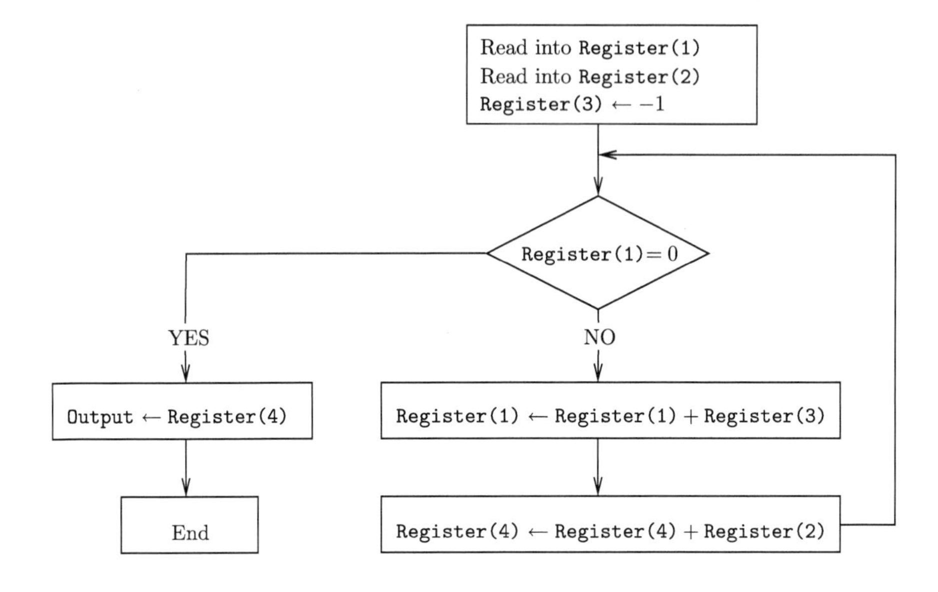

Fig. 2.9

$$\underbrace{b + b + b + \ldots + b}_{a \text{ times}},$$

i.e., to sum a many values b.

Exercise 2.12 Assume $a = 3$ and $b = 7$. Execute the computation of the program on this input. Depict a table that shows the content of all registers after particular steps.

If $a = 0$, then the result has to be $a \cdot b = 0$. The programs work correctly for $a = 0$, because a is read into `Register(1)` and the test in row 4 leads directly to row 8, in which the value 0 as the content of `Register(4)` is displayed.
If $a \geq 0$, the execution of the instruction in row 6 causes that b is added to the content `Register(4)` that has to contain the final result at the end of the computation. The execution of the instruction in row 5 results in a decrease of the content of `Register(1)` by 1. At the beginning `Register(1)` contained a. After the i-th execution of the loop (see Fig. 2.9) for an $i < a$, `Register(1)` contains $a - i$ and `Register(4)` contains the number

$$\underbrace{b + b + \ldots + b}_{i \text{ times}} = i \cdot b.$$

If "`Register(1) = 0`", we know that the loop was executed exactly a times and so `Register(4)` contains the value

$$\underbrace{b + b + b + \ldots + b}_{a \text{ times}} = a \cdot b.$$

In this way we developed a program that can multiply two integers without using the operation of multiplication. This means that removing multiplication from the list of our basic instructions does not decrease the power of our algorithms and so does not affect our notion of "algorithm".

But the program in Fig. 2.9 has a drawback. At the beginning, we said that a and b are integers. What does the program do if a or b is negative? If b is negative and a is positive, then the program works orderly. But if a is negative, the content of `Register(1)` will never[5] be 0 and so the loop will be repeated infinitely many times. □

Exercise 2.13 How do we proceed in order to modify the program in Fig. 2.9 to a program that correctly multiplies two integers a and b, also when a is negative?

Exercise 2.14 Try to write a program that computes the sum $a + b$ for two given natural numbers a and b and that uses only the new arithmetic instructions

```
Register(i) ← Register(i)+1
Register(j) ← Register(j)-1,
```

which increase respectively decrease the content of a register by 1. All other arithmetic operations are not allowed and the only allowed test operation is the question whether the content of a register is 0 or not.

Finally, one can see that all algorithms can be implemented as programs that use the test on 0, addition by 1, subtraction by 1, and some input/output instructions only. Therefore, there is no doubt about the automatic executability of algorithms.

[5] At the beginning `Register (1)` gets the negative value a that is only going to be decreased during the run of the program.

Only for those who want to know the whole list of basic computer instructions do we present more details. First of all, computing the root of a number is not a basic instruction. To compute the root of an integer, one has to write a program that uses the arithmetic operations $+, -, *$, and $/$ only. Since this requires a non-negligible effort we avoid developing such a program here.

On the other hand, some indispensable basic instructions are still missing. To see this, consider the following task. The input is a sequence of integers. We do not know how many there are. We only recognize the end of the sequence by reading 0, which can be viewed as the endmaker of the sequence. The task is only to read all integers of the sequence and to save all of them in `Register(101)`, `Register(102)`, `Register(103)`, etc., one after another. This saving is finished when the last integer read is 0. One could start to design a program as follows:

```
1 Read into Register(1)
2 If Register(1) = 0, then go to row □
3 Register(101) ← Register(1)
4 Read into Register(1)
5 If Register(1) = 0, then go to row □
6 Register(102) ← Register(1)
7 Read into Register(1)
8 If Register(1) = 0, then go to row □
9 Register(103) ← Register(1)
  ⋮
```

We always read the next number into `Register(1)`, and if the number is different from 0 then we save it in the first free register after `Register (101)`. The problem is that we do not know how to continue writing the program. If the input queue contains 17, −6, and 0, then we are already done. If the queue contains 1000 integers different from 0, then this program has to have 3000 rows. But we do not know when to stop writing the program and so we do not know where to put the row with the instruction `end`. We used the notation □ in the program, because we did not know where to put `end`. Certainly we are not allowed to write an infinite program. A common idea in similar situations is to use a loop. One could try to design a loop such as that in Fig. 2.10.

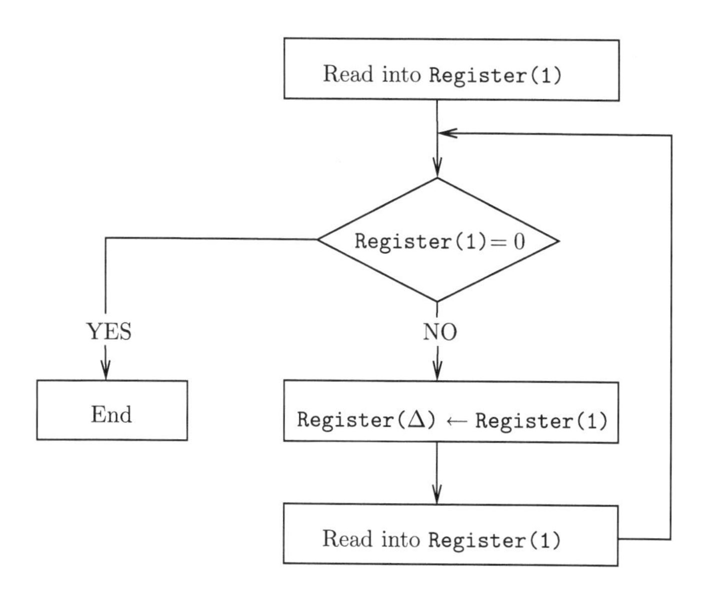

Fig. 2.10

The only problem is that we do not know in which `Register(△)` the actual integer has to be saved. Clearly, we cannot always use the same integer △, because we want to save all integers. We know that we want to save the integer at the address $100 + i$ in the i-th run of the loop. But we are not able to do this, because our instructions allow us to use a fixed address for △ in `Register(△)`.

Therefore, we introduce new instructions that use so-called indirect addressing. The instruction

(14) `Register(Register(`i`))` $\leftarrow$ `Register(`j`)`

for positive integers i and j means that the content of `Register(`j`)` has to be saved in the register, whose address is the content of `Register(`i`)`.

Is this confusing? Let us explain it transparently using an example. Assume that the content of `Register(3)` is 112 and that `Register(7)` contains 24. The computer has to execute the instruction

`Register(Register(3))` $\leftarrow$ `Register(7)`.

First, the computer takes the content of `Register(3)` and sees that it is 112. Then the computer executes the already known instruction

`Register(112)` $\leftarrow$ `Register(7)`.

In this way, the number 24 (the content of `Register(7)`) is saved in `Register(112)`. The contents of all registers except for `Register(112)` remain unchanged.

Exercise 2.15 Most of the computer instructions introduced have a version with indirect addressing. Try to explain the meanings of the following instructions!

a) `Register(`k`)` $\leftarrow$ `Register(Register(`m`))`
b) `Register(Register(`i`))` $\leftarrow$ `Register(`l`)*Register(`j`)`

Using indirect addressing one can solve our problem of saving data of unknown number as depicted in Fig. 2.11. We use `Register(2)` for saving the address at which the next integer has to be saved. At the beginning, we put 101 into `Register(2)`, and then, after saving the next integer, we increase the content of `Register(2)` by 1. The number 1 lies in `Register(3)` during the whole computation.

Exercise 2.16 The input queue contains the integer sequence $113, -7, 20, 8, 0$. Simulate step by step the work of the program in Fig. 2.11 on this input! Determine the contents of the registers with the addresses $1, 2, 3, 100, 101, 102, 103, 104$, and 105 after each step! Assume that at the beginning all registers contain 0.

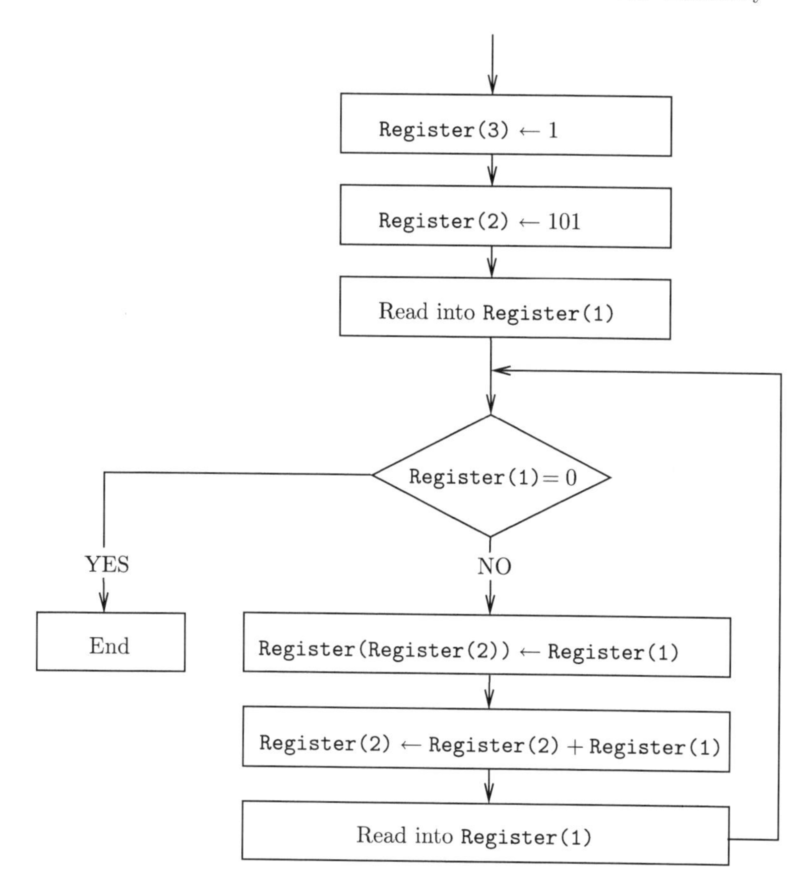

Fig. 2.11

2.5 Summary or What We Have Learnt Here

It does not matter whether you believe it or not, if you were able to solve a few of the exercises formulated above then you have already programmed and so you have learnt a little bit about what it means to work as a programmer. But this was not the main goal of this chapter.

Our aim was to explain the meaning of the notion of an algorithm. We understand that our expectation on the definition of the notion of an algorithm as a formalization of the notion of a method corresponds to the following requirements:

1. One has to be able to apply an algorithm (a method) even if one is not an expert in solving the considered problem. One does not need to understand why the algorithm provides the solution of the problem. It is sufficient to be able to execute the simple activities the algorithm consists of. Defining the notion of an algorithm, one has to list all such simple activities and everybody has to agree that all these activities are executable by a machine.

2. An algorithm is designed not only to solve a problem instance, but it must be applicable to solving all possible instances of a given problem. (Please be reminded that a problem is a general task such as sorting or solving quadratic equations. A problem instance corresponds to a concrete input such as "Sort the integer sequence $1, 7, 3, 2, 8$" or "Solve the quadratic equation $2x^2 - 3x + 5 = 0$".)

3. We require a guarantee that an algorithm for a problem successfully finds a solution for each problem instance. This means that the algorithm always finishes its work in a finite time and its output corresponds to a correct solution to the given input.

An algorithm can be implemented as a program in a programming language. A program is an algorithm representation that is understandable for the computer. But a program is not a synonym of the notion of an algorithm. A program is only a sequence of instructions that are understandable and executable by a computer. This instruction sequence does not necessarily lead to a reasonable activity. For instance, the execution of a program can lead to some pointless computations that do not solve any problem or to an infinite work in a loop.

Solutions to Some Exercises

Exercise 2.2 A cooking algorithm for heating 1 l of water to 90°C can be described as follows:

1. Pour 1 l water into the pot T.
2. Put the pot T on the hotplate for 15 seconds and then take it away.

3. If the temperature of the water is at least 90°C, finish the work! Else continue with step 2.

Exercise 2.3 The development of memory can be transparently represented using the following table:

	1	2	3	4	5
Input queue	11, 12, 13	12, 13	12, 13	13	
Register(0)	1	2	3	4	5
Register(1)	0	11	11	11	11
Register(2)	1117	1117	100	100	13
Register(3)	21	21	21	12	12
Register(4)	0	0	0	0	0

The first column represents the state of the memory before the execution of the program started. The $(i+1)$-th column describes the solution immediately after the execution of the i-th instruction and so before the execution of the $(i+1)$-th instruction.

Exercise 2.8 The given program reads the three integers from the input queue (rows 1, 2, and 3). Then it computes their sum and saves it in `Register(4)` (program rows 4 and 5). In the program rows 6 and 7 the average of the input integers is computed and saved in `Register(6)`. The instruction in row 8 displays the average value. The following table shows the development of the computer states after particular steps. In contrast to the table in Exercise 2.3, we write the values in the table only when the content of a register has changed in the previous step.

Input	a, b, c	b, c	c							
Register(0)	1	2	3	4	5	6	7	8	9	10
Register(1)	0	a								
Register(2)	0		b							
Register(3)	0			c						
Register(4)	0				$a+b$	$a+b+c$				
Register(5)	0						3			
Register(6)	0							$\frac{a+b+c}{3}$		
Output									$\frac{a+b+c}{3}$	

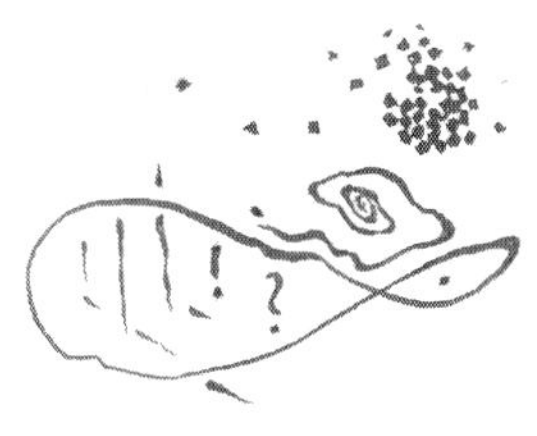

Little progress would be made in the world
if we were always afraid of possible negative consequences.

Georg Christoph Lichtenberg

Chapter 3

Infinity Is Not Equal to Infinity, or Why Infinity Is Infinitely Important in Computer Science

3.1 Why Do We Need Infinity?

The known universe is finite, and most physical theories consider the world to be finite. Everything we see and each object we touch is finite. Whatever we do in reality, we come in contact with finite things only.

> *Why then deal with infinity? Is infinity not something artificial, simply a toy of mathematics?*

In spite of possible doubts that may appear when we meet the concept of infinity for the first time, we claim that infinity is an unavoidable instrument for the successful investigation of our finite world. We touch infinity for the first time in elementary school, where we meet the set

J. Hromkovič, *Algorithmic Adventures*, DOI 10.1007/978-3-540-85986-4_3,

$$\mathbb{N} = \{0, 1, 2, 3, \ldots\}$$

of natural numbers (nonnegative integers). The concept of this set can be formulated as follows:

> For each natural number i, there is a larger natural number $i+1$.

In other words, there does not exist any number that is larger than all other numbers (i.e., there exists no largest number), because for each number x there are numbers larger than x. What is the consequence of this concept? We are unable to write down the list of all natural numbers. It does not matter how many of them we have written already, there are still many missing. Hence, our writing is a never-ending story, and because of this we speak about **potential infinity** or about an **unbounded** number of natural numbers. We have a similar situation with the idea (the notion) of a line in geometry. Any line is potentially infinite, and so its length is unbounded (infinitely large). One can walk along a line for an arbitrarily long time and one never reaches the end; it does not matter which point (position) of the line you have reached, you can always continue to walk further in the same direction.

The main trouble with understanding the concept of infinity is that we are not capable of imagining any infinite object at once. We simply cannot see **actual infinity**. We realize that we have infinitely (unboundedly) many natural numbers, but we are not able to see all natural numbers at once. Similarly we are unable to see a whole infinite line at once. We are only able to see a finite fraction (part) of an infinite object. The way out we use is to denote infinite objects by symbols and then to work with these symbols as finite representations of the corresponding infinite objects.

To omit infinity, one can propose exchanging unbounded sizes with a huge finite bound. For instance, one can take the number[1] of all protons in the Universe as the largest number and forbid all larger numbers. For most calculations and considerations one can be successful with this strategy. But not if you try to compute

[1] This number consists of 79 decimal digits.

the whole energy of the Universe or if you want to investigate all possible relations between the particles of the Universe. It does not matter what huge number one chooses as the largest number allowed, there appear reasonable situations whose investigation requires us to perform calculations with numbers larger than the upper bound proposed. Moreover, for every number x, we are not only aware of the existence of a number larger than x, we are even able to write this larger number down and see it as a concrete object. Why should we forbid something we can imagine (and thus has a concrete representation in our mind) and that we may even need?

To convince the reader of the usefulness of the concept of infinity, we need to provide more arguments than presenting the natural existence of potential infinity. We claim that by means of the concept of infinity we are able to investigate the world more successfully than without, and so that infinity contributes to a better understanding of the finite world around. Infinity does more than enable us to deal with infinitely large sizes; we can also consider infinitely small sizes.

> *What is the smallest positive rational number, i.e., what is the smallest positive fraction larger than 0?*

Consider the fraction 1/1000. We can halve it and get the fraction 1/2000, which is smaller than 1/1000. Now we can halve the resulting fraction again and get 1/4000 . . . . It does not matter which small positive fraction

$$\frac{1}{x}$$

one takes, by halving it one gets the positive fraction

$$\frac{1}{2x}.$$

This fraction $1/2x$ is smaller than $1/x$ and surely still larger than 0. We see that this procedure of creating smaller and smaller numbers does not have any end too. For each positive number, there exists a smaller positive number, etc.

David Hilbert (1862–1943), one of the most famous mathematicians, said:

> *"In some sense, the mathematical analysis is nothing other than a symphony about the topic of infinity."*

We add to this quotation that current physics as we know it would not exist without the notion of infinity. The key concepts and notions of mathematics such as derivation, limit, integral and differential equations would not exist without infinity. How can physics model the world without these notions? It is unimaginable. One would already have trouble in building fundamental notions of physics. How can one define acceleration without these mathematical concepts? Many of the notions and concepts of mathematics were created because physics had a strong need to introduce and to use them.

The conclusion is that large parts of mathematics would disappear if infinity were forbidden. Since mathematics is the formal language of science, and we often measure a degree of "maturity" of scientific disciplines with respect to use this language, the exclusion of the notion of infinity would set science back several hundred years.

We have the same situation in computer science where we have to distinguish between programs, which allow infinite computations, and algorithms, which guarantee a finite computation on each input. There are infinitely many programs and infinitely many algorithmic tasks. A typical computing problem consists of infinitely many problem instances. Infinity is everywhere in computer science, and so computer scientists cannot live without this concept.

The goal of this chapter is not only to show that the concept of infinity is a research instrument of computer science. Our effort will be strange because we do not satisfy ourselves with trouble that appears when we are dealing with potential infinity and actual infinity (which no one has ever seen). We will still continue to pose the following strange question:

Does there exist only one infinity or do there exist several differently large infinities?

Dealing with this question, which seems to be stupid and too abstract at first, was and is of enormous usefulness for science. Here we follow some of the most important discoveries about infinity in order to show that there exist at least two differently sized infinities. What is the gain of this? We can use this knowledge to show that the number of algorithmic problems (computing tasks) is larger than the number of all programs. In this way we obtain the first fundamental discovery of computer science.

One cannot automate everything. There are tasks for which no algorithm exists and so which cannot be automatically solved by any computer or robot.

As a result of this discovery we are able to present in the next chapter concrete problems from practice that are not algorithmically (automatically) solvable. This is a wonderful example showing how the concept of an object that does not exist in the real world can help to achieve results and discoveries that are of practical importance. Remember, using hypothetical and abstract objects in research is typical rather than exceptional. And the most important thing is whether the research goal was achieved. Success is the measure of usefulness of new concepts.

3.2 Cantor's Concept for Comparing the Sizes of Infinite Sets

Comparing finite numbers is simple. All numbers lie on the real axis in increasing order from left to right. The smaller of two numbers is always to the left of the other one (Fig. 3.1).

Fig. 3.1

Hence, 2 is smaller than 7 because it is to the left of 7 on the axis. But this is not a concept for comparing numbers because the numbers are a priori positioned on the axes in such a way that they increase from left to right and decrease from right to left. Though the axis is infinite in both directions, only finite numbers lie on it. It does not matter which position (which point) we consider, the number sitting there is always a concrete finite number. This is the concept of potential infinity. One can move along the axis arbitrarily far to the right or to the left, and each position reached on this trip contains a concrete finite number. There are no infinite numbers on the axis. To denote infinity in mathematics we use the symbol

$$\infty$$

called a "laying eight". Originally this symbol came from the letter aleph of the Hebrew alphabet. But if one represents infinity by just one symbol ∞, there does not exist any possibility of comparing different infinities.

What do we need to overcome this?

We need a new representation of numbers. To get it, we need the notion of a set. A set is any collection of objects (elements) that are pairwise distinct. For instance, $\{2, 3, 7\}$ is a set that contains three numbers $2, 3$, and 7. The set {John, Anna, Peter, Paula} contains four objects (elements): John, Anna, Peter, and Paula. For any set A, we use the notation

$$|A|$$

for the number of elements in A and call $|A|$ the **cardinality (size) of *A***. For instance,

$$|\{2, 3, 7\}| = 3, \text{ and } |\{\text{John, Anna, Peter, Paula}\}| = 4 .$$

Now, we take the sizes of sets as representations of numbers. In this way the cardinality of the set $\{2, 3, 7\}$ represents the integer 3, and the cardinality of the set {John, Anna, Peter, Paula} represents the number 4. Clearly, every positive integer gets many different representations in this way. For instance

$$|\{1, 2\}| , |\{7, 11\}| , |\{\text{Petra, Paula}\}| , |\{\square, \bigcirc\}|$$

are all representations of the integer 2. Is this not fussy? What is the advantage of this seemingly too complicated representation of integers?

Maybe you find this representation to be awkward for the comparison of finite numbers.[2] But, by using this way of representing numbers, we gain the ability to compare infinite sizes. The cardinality

$$|\mathbb{N}|$$

for $\mathbb{N} = \{0, 1, 2, \ldots\}$ is the infinite number that corresponds to the number of all natural numbers. If $\mathbb{Q}^+$ denotes the set of all positive rational numbers, then the number

$$|\mathbb{Q}^+|$$

represents the infinite number that corresponds to the number of all positive rational numbers (fractions). And

$$|\mathbb{R}|$$

is the infinite number that corresponds to the number of all real numbers, assuming $\mathbb{R}$ denotes the set of all real numbers. Now we see the advantage. We are allowed to ask

"Is $|\mathbb{N}|$ smaller than $|\mathbb{R}|$?"

or

"Is $|\mathbb{Q}^+|$ smaller than $|\mathbb{R}|$?"

As a result of representing numbers this way we are now able to pose the question whether an infinity is larger than another infinity.

We have reduced our problem of comparing (infinite) numbers to comparing sizes of (infinite) sets. But now the following question arises:

[2] With high probability, this is the original representation of natural numbers used by Stone Age men. Small children use first the representation of numbers by sets in order to later develop an abstract concept of a "number".

How to compare the sizes of two sets?

If the sets are finite, then the comparison is simple. One simply counts the number of elements in both sets and compares the corresponding cardinalities. For sure, we cannot do this for infinite sets. If one tried to count the elements of infinite sets, then the counting would never end, and so the proper comparison would never be performed. Hence, we need a general method for comparing sizes of sets that would work for finite as well as infinite sets and that one could judge as reasonable and trustworthy. This means that we are again at the deepest axiomatic level of science. Our fundamental task is to create the notion of infinity and the definition of **"smaller than or equal to"** for the comparison of the cardinalities of two sets.

Now we let a shepherd help us. This is no shame because mathematicians did the same.

Fig. 3.2

Fig. 3.3

A shepherd has a large flock of sheep with many black and white sheep. He never went to school, and though he is wise (which means that he cannot leave the mountains), he can count only to five. He wants to find out whether he has more black sheep than white ones or vice versa (Fig. 3.2).

How can he do it without counting? In the following simple and ingenious way. He simply takes one black sheep and one white sheep and creates one pair

(white sheep, black sheep),

and sends them away from the flock. Then he creates another white–black pair and sends it away too (Fig. 3.3). He continues in this way until he has sheep of one color only or there are no remaining sheep at all (i.e., until there is no way to build a white–black pair of sheep). Now he can reach the following conclusion:

(i) If no sheep remained, he has as many white sheep as black ones.

(ii) If one or more white sheep remained in the flock, then he has more white sheep than black ones (Fig. 3.3).

(iii) If one or more black sheep remained in the flock, then he has more black sheep than white ones.

Pairing the sheep and conclusion (i) is used by mathematicians as the basic for comparing the sizes of sets.

Definition 3.1. Let A and B be two sets. A **matching** of A and B is a set of pairs (a, b) that satisfies the following rules:

(i) Element a belongs to A ($a \in A$), and element b belongs to B ($b \in B$).

(ii) Each element of A is the first element of exactly one pair (i.e., no element of A is involved in two or more pairs and no element of A remains unmatched).

(iii) Each element of B is the second element of exactly one pair.

For each pair (a, b), we say that ***a* and *b* are married**. We say that **A and B have the same size** or that **the size of A equals the size of B** and write

$$|\boldsymbol{A}| = |\boldsymbol{B}|$$

if there exists a matching of A and B. We say that **the size of A is not equal to the size of B** and write

$$|\boldsymbol{A}| \neq |\boldsymbol{B}|$$

if there does not exist any matching of A and B.

Consider the two sets $A = \{2, 3, 4, 5\}$ and $B = \{2, 5, 7, 11\}$ depicted in Fig. 3.4. Figure 3.4 depicts the matching

$$(2, 2), (3, 5), (4, 7), (5, 11) .$$

Each element of A is involved in exactly one pair of the matching as the first element. For instance, the element 4 of A is involved as the first element of the third pair $(4, 7)$. Each element of B is involved in exactly one pair as the second element. For instance, the element 5 of B is involved in the second pair. In other words, each element of A is married to exactly one element of B, each

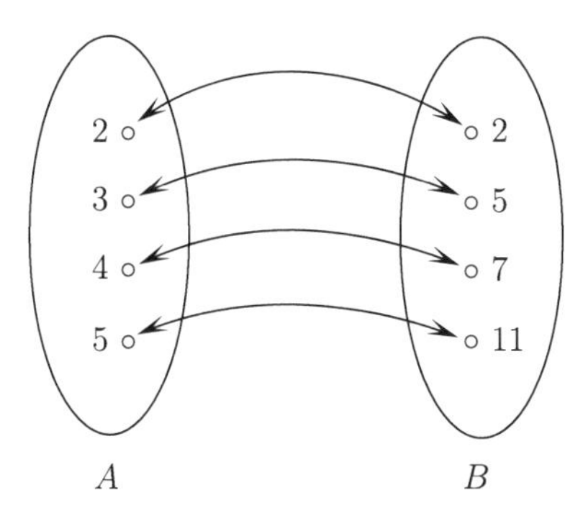

Fig. 3.4

element of B is married to exactly one element of A, and so no element of A or B remains single. Therefore, we can conclude

$$|\{2, 3, 4, 5\}| = |\{2, 5, 7, 11\}| \ .$$

You can also find other matchings of A and B. For instance,

$$(2, 11), (3, 7), (4, 5), (5, 2)$$

is also a matching of A and B.

Exercise 3.1 (a) Give two other matchings of the sets $A = \{2, 3, 4, 5\}$ and $B = \{2, 5, 7, 11\}$.
(b) Why is $(2, 2), (4, 5), (5, 11), (2, 7)$ not a matching of A and B?

Following this concept of comparing the sizes of two sets, a set A of men and a set B of women are equally sized if all the women and men from A and B can get married in such a way that no single remains.[3]

A matching of the sets $C = \{1, 2, 3\}$ and $D = \{2, 4, 6, 8\}$ cannot exist because every attempt to match the elements of D and C ends in the situation where one element of D remains single. Therefore, $|D| \neq |C|$ holds. An unsuccessful attempt to match C and D is depicted in Fig. 3.5.

[3] Building same-sex pairs is not allowed here.

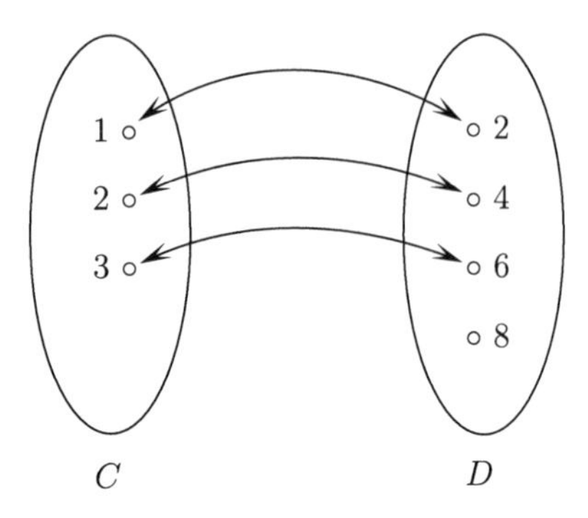

Fig. 3.5

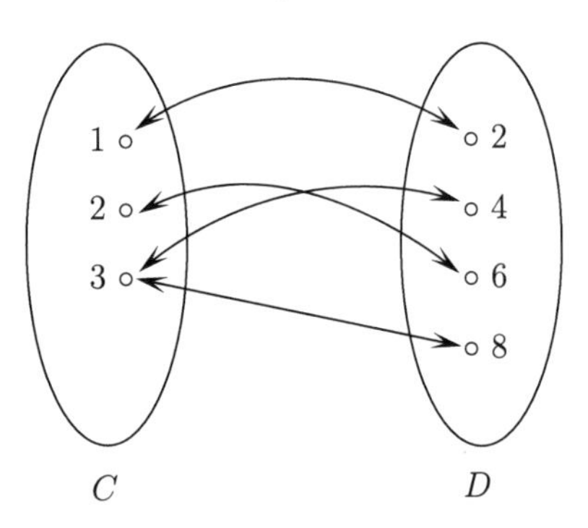

Fig. 3.6

Figure 3.6 shows another attempt to match C and D. Here the result is not a matching of C and D because element 3 of C is married to two elements 4 and 8 of D.

But we do not need the concept of matching in order to compare the sizes of finite sets. We were also able to do it without this concept. In the previous description, we only checked that our matching concept works in the finite world.[4] In what follows we try to apply this concept to infinite sets. Consider the two sets

$$\mathbb{N}_{even} = \{0, 2, 4, 6, 8, \ldots\}$$

of all even natural numbers and

$$\mathbb{N}_{odd} = \{1, 3, 5, 7, 9, \ldots\}$$

[4] If the concept did not work in the finite world, then we would have to reject it.

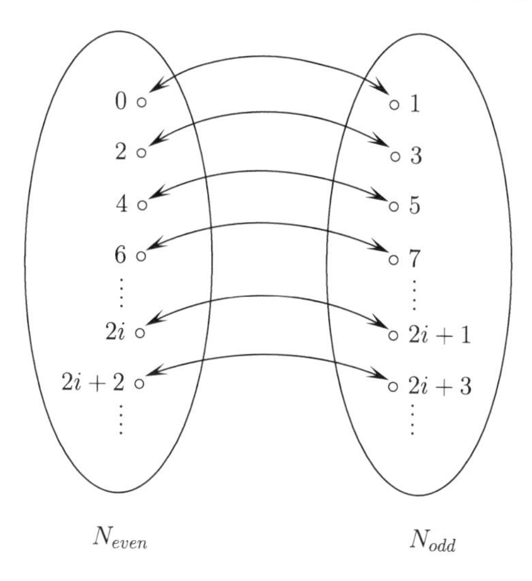

Fig. 3.7

of all odd natural numbers. At first glance, these sets look to be of the same size, and we try to verify this by means of our concept. We match each even number $2i$ to the odd number $2i + 1$.

Following Fig. 3.7, we see that we get an infinite sequence of pairs

$$(0, 1), (2, 3), (4, 5), (6, 7), \ldots, (2i, 2i + 1), \ldots$$

in this way. This sequence of pairs is a correct matching of A and B. No element from $\mathbb{N}_{even}$ or of $\mathbb{N}_{odd}$ is involved in two or more pairs (is married to more than one element). On the other hand no element remains single (unmarried). For each even number $2k$ from $\mathbb{N}_{even}$, we have the pair $(2k, 2k + 1)$. For each odd number $2m+1$ from $\mathbb{N}_{odd}$, we have the pair $(2m, 2m+1)$. Hence, we verified that the equality $|\mathbb{N}_{even}| = |\mathbb{N}_{odd}|$ holds.

Exercise 3.2 Prove that $|\mathbb{Z}^+| = |\mathbb{Z}^-|$, where $\mathbb{Z}^+ = \{1, 2, 3, 4, \ldots\}$ and $\mathbb{Z}^- = \{-1, -2, -3, -4, \ldots\}$. Draw a figure depicting your matching as we did for $\mathbb{N}_{even}$ and $\mathbb{N}_{odd}$ in Fig. 3.7.

Up to this point everything looks tidy, understandable, and acceptable. Now, we present something which may be difficult to

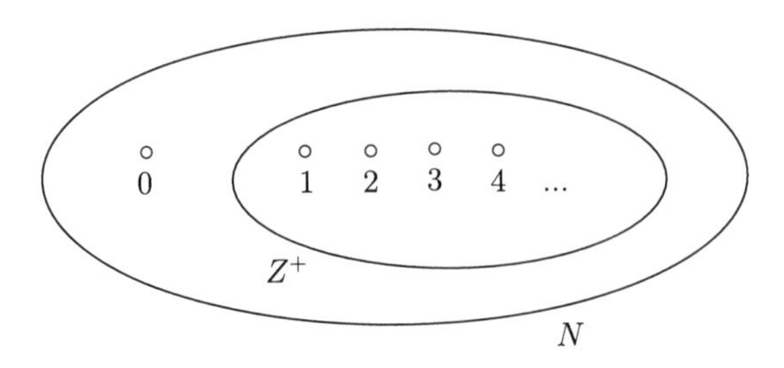

Fig. 3.8

come to terms with, at least at the first attempt. Consider the sets

$$\mathbb{N} = \{0, 1, 2, 3, \ldots\} \text{ and } \mathbb{Z}^+ = \{1, 2, 3, 4, \ldots\} \ .$$

All elements of $\mathbb{Z}^+$ are in $\mathbb{N}$, and so

$$\mathbb{Z}^+ \subseteq \mathbb{N} \ ,$$

i.e., $\mathbb{Z}^+$ is a **subset** of $\mathbb{N}$. Moreover, the element 0 belongs to $\mathbb{N}$ ($0 \in \mathbb{N}$), but not to $\mathbb{Z}^+$ ($0 \notin \mathbb{Z}^+$). We therefore say that $\mathbb{Z}^+$ is a **proper subset** of $\mathbb{N}$ and write $\mathbb{Z}^+ \subset \mathbb{N}$. The notion "$A$ is a proper subset of B" means that A is a part of B but not the whole of B. We can see this situation transparently for the case

$$\mathbb{Z}^+ \subset \mathbb{N}$$

in Fig. 3.8. We see that $\mathbb{Z}^+$ is completely contained in $\mathbb{N}$ but $\mathbb{Z}^+$ does not cover the whole of $\mathbb{N}$ because $0 \in \mathbb{N}$ and $0 \notin \mathbb{Z}^+$.

However, we claim that

$$|\mathbb{N}| = |\mathbb{Z}^+|$$

is true, i.e., that the sizes of $|\mathbb{N}|$ and $|\mathbb{Z}^+|$ are equal. We justify this claim by building the following matching

$$(0, 1), (1, 2), (2, 3), \ldots, (i, i + 1), \ldots \ ,$$

depicted in Fig. 3.9.

We clearly see that all elements of $\mathbb{N}$ and $\mathbb{Z}^+$ are correctly married. No element remains single. The conclusion is that $\mathbb{N}$ is not larger

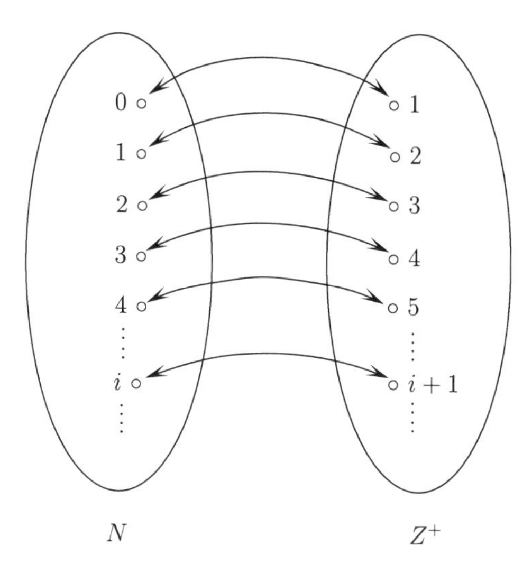

Fig. 3.9

than $\mathbb{Z}^+$ though $\mathbb{N}$ has one more element than $\mathbb{Z}^+$. But this fact may not be too surprising or even worrying. It only says that

$$\infty + 1 = \infty ,$$

and so that increasing infinity by 1 does not lead to a larger infinity. This does not look surprising. What is 1 in comparison with infinity? It is nothing and can be neglected. This at first glance surprising combination of the facts

$$\mathbb{Z}^+ \subset \mathbb{N} \text{ (Fig. 3.8) and } \left|\mathbb{Z}^+\right| = |\mathbb{N}| \text{ (Fig. 3.9)}$$

provides the fundamentals used for creating the mathematical definition of infinity. Mathematicians took thousands of years to find this definition and then a new generation of researchers was needed to be able to accept it and fully imagine its meaning. It was not so easy for these mathematicians to see that this definition provides what they strived for, namely a formal criterion for distinguishing between finite sets and infinite sets.

Definition 3.2. A set A is infinite if and only if there exists a proper subset B of A such that

$$|A| = |B| \ .$$

In other words:

> *An object is* ***infinite*** *if there is a proper part of the object that is as large as the whole object.*

Now you can say: *"Stop! This is too much for me. I cannot accept something like that. How can a part be of the same size as the whole? Something like this does not exist."*

It is excellent that you have this opinion. Especially because of this, this definition is good. In the real world in which everything is finite, no part can be as large as the whole. This is exactly what we can agree on. No finite (real) object can have this strange property. And, in this way, Definition 3.2 says correctly that all such objects are finite (i.e., not infinite). But in the artificial world of infinity, it is not only possible to have this property, but also necessary. And so this property is exactly what we were searching for, since a thing that has this property is infinite and one that does not have this property is finite. In this way, Definition 3.2 provides a criterion for classifying objects into finite and infinite and this is exactly what one expects from such a definition.

To get a deeper understanding of this at first glance strange property of infinite objects, we present two examples.

Example 3.1 Hotel Hilbert

Let us consider a hotel with infinitely many single rooms that is known as the Hotel Hilbert. The rooms are enumerated as follows:

$$Z(0), Z(1), Z(2), Z(3), \ldots, Z(i), \ldots \quad .$$

All rooms are occupied, i.e., there is exactly one guest in each room. Now, a new guest enters the hotel and asks the porter: "Do you have a free room for me?" "No problem", answers the porter and accommodates the new guest by the following strategy. He asks every guest in the hotel to move to the next room with the number that is 1 higher than the room number of the room used up till now. Following this request, the guest in room $Z(0)$ moves to the room $Z(1)$, the guest in $Z(1)$ moves to $Z(2)$, etc. In general, the guest in $Z(i)$ moves to the room $Z(i+1)$. In this way, the room

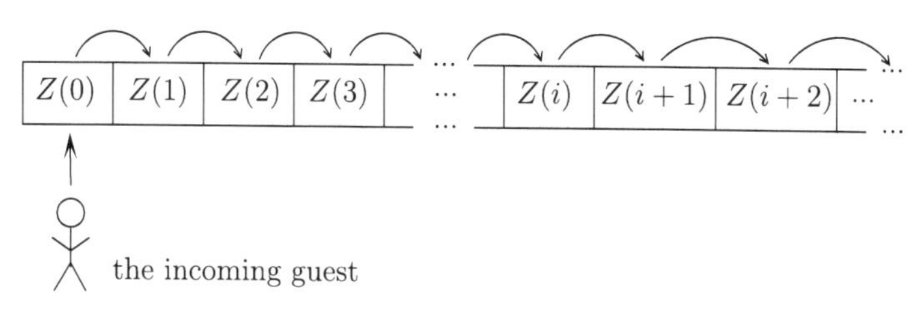

Fig. 3.10

$Z(0)$ becomes free, and so $Z(0)$ can be assigned to the newcomer (Fig. 3.10).

We observe that, after the move, every guest has her or his own room and room $Z(0)$ becomes free for the newcomer. Mathematicians argue for the truthfulness of this observation as follows. Clearly room $Z(0)$ is free after the move. The task is to show that every guest has his or her own room after the move. Let G be an arbitrary guest. This person G stays alone in a specific room before the move. Let $Z(n)$ be the number of this room. Following the instructions of the porter, guest G moves from $Z(n)$ to $Z(n+1)$. He can do this because $Z(n+1)$ becomes free because the guest in this room moved to room $Z(n+2)$. Hence, after the moves guest G stays alone in room $Z(n+1)$. Since our argument is valid for every guest of the hotel, all guests have a single room accommodation after the move.

The solution above shows why the real infinity was considered as a paradox[5] of mathematics for a long time. Hotel Hilbert is a real infinity. Something like this can only be outlined by drawing a finite part of it and adding $\cdots$. But nobody can see it at once. Hence, it is not possible to observe the whole move of infinitely many guests at once. On the other hand, observing each particular guest separately, one can verify that the move works successfully.

Only when one was able to realize that infinity differs from finiteness by having proper subparts of the same size as the whole, was this paradox solved[6]. We observe that the move corresponds to

[5] a seemingly contradictory fact or an inexplicable situation
[6] and so it is not a paradox anymore

matching the elements of the set $\mathbb{N}$ (the set of guests) with the set $\mathbb{Z}^+$ (the set of rooms up to room $Z(1)$). □

Exercise 3.3 (a) Three newcomers enter Hotel Hilbert. As usual, the hotel is completely booked. Play the role of the porter and accommodate the three new guests in such a way that no former guest has to leave the hotel, and after the move each new guest and each former guest possesses their own room. If possible, arrange the accommodation using one move of each guest instead of organizing 3 moves one after each other.
(b) A newcomer enters Hotel Hilbert and asks for his favored room, $Z(7)$. How can the porter satisfy this request?

We take the next example from physics. Physicists discovered it as a remedy for depression caused by imagining that our Earth and mankind are tiny in comparison with the huge Universe[7].

Example 3.2 Let us view our Earth and Universe as infinite sets of points of size 0 that can lie arbitrarily close each to each other. To simplify our story we view everything two-dimensionally instead of working in three dimensions. The whole Universe can be viewed as a large sheet of paper, and Earth can be depicted as a small circle on the sheet (Fig. 3.11). If somebody has doubts about viewing our small Earth as an infinite set of points, remember that there are infinitely many points on the finite part of the real axis between the numbers 0 and 1. Each rational number between 0 and 1 can be viewed as a point on the line between 0 and 1. And there are infinitely many rational numbers between 0 and 1. We proved this fact already by generating infinitely many rational numbers between 0 and 1 in our unsuccessful attempt to find the smallest positive rational number.

Another justification of this fact is related to the proof of the following claim.

> For any two different rational numbers a and b, $a < b$, there are infinitely many rational numbers between a and b.

The first number between a and b we generate is the number $c_1 = \frac{a+b}{2}$, i.e., the average value of a and b. The next one is $c_2 = \frac{c_1+b}{2}$,

---

[7] In this way, physicists try to ease the negative consequences of their discoveries.

i.e., the average of c_1 and b. In general, the i -th generated number from $[a, b]$ is

$$c_i = \frac{c_{i-1} + b}{2},$$

i.e., the average of c_{i-1} and b. When $a = 0$ and $b = 1$, then one gets the infinite sequence

$$\frac{1}{2}, \frac{3}{4}, \frac{7}{8}, \frac{15}{16}, \ldots$$

of pairwise different rational numbers between 0 and 1.

Now let us finally switch to the fact physicists want to tell us. All points of our huge Universe beyond Earth can be matched with the points of Earth. This claim has two positive (healing) interpretations:

(i) The number of points of our Earth is equal to the number of points of the Universe outside Earth.

(ii) Everything that happens in the Universe can be reflected on Earth and so can be imitated in our tiny world.

Hence, our task is to search for a matching between the Earth points and the points outside Earth. In what follows we show how to assign an Earth point P_E to any point P_U outside Earth.

First, we connect P_U and the Earth center M by a line (Fig. 3.11). The point P_E we are searching for has to lie on this line. Next, we depict the two tangents t_1 and t_2 of the circle that goes through the point P_U (Fig. 3.11). Remember that a tangent of a circle is a line that has exactly one common point with the circle. We call the point at which t_1 touches the circle A_P and we denote by B_P the common point[8] of the circle and the line t_2 (see Fig. 3.11). Finally, we connect the points B_P and A_P by a line $B_P A_P$ (Fig. 3.12). The point at the intersection of the lines $B_P A_P$ and $P_U M$ is the Earth point P_E we assign to P_U (Fig. 3.12).

[8] Mathematicians would say that the point A_P is the intersection of the circle and t_1 and that B_P is the intersection of the circle and t_2.

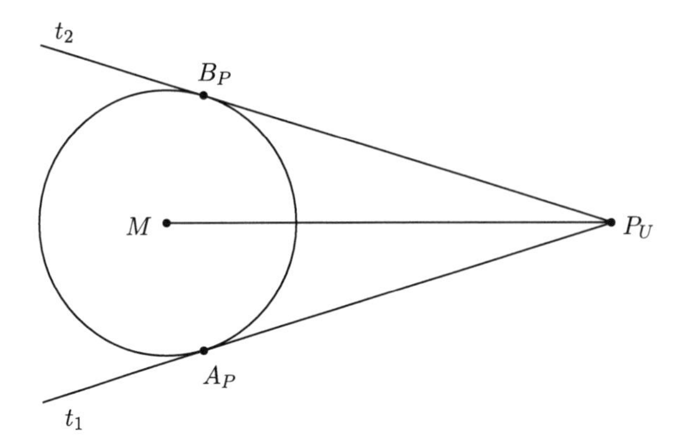

Fig. 3.11

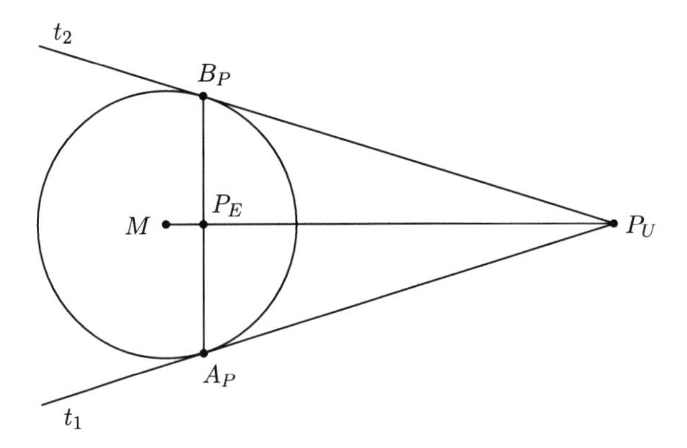

Fig. 3.12

Next, we have to show that this geometric assignment of P_E to P_U defines a matching between the Earth's points and the points outside Earth. Namely we have to show that one always assigns two distinct Earth points P_E and P'_E to two different points P_U and P'_U outside Earth.

To verify this fact, we distinguish two possibilities with respect to the positions of P_U and P'_U according to M.

(i) The points M, P_U, and P'_U do not lie on the same line. This situation is depicted in Fig. 3.13. We know that P_E lies on the line MP_U and that P'_E lies on the line MP'_U. Since the only common point of the lines MP_U and MP'_U is M, and M is different from P_E and P'_E, independently of the positions of P_E and P'_E on their lines, the points P_E and P'_E must be different.

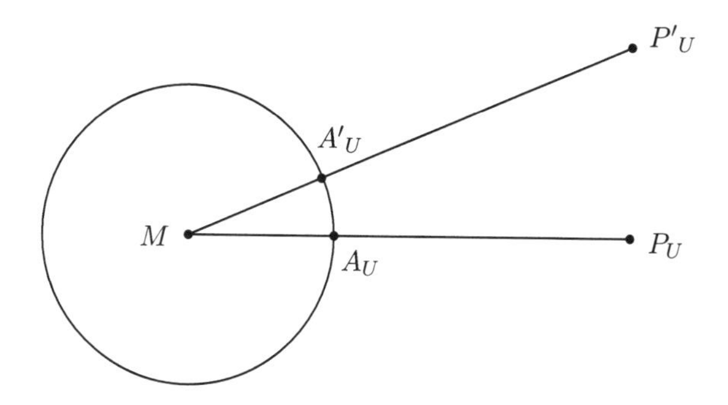

Fig. 3.13: E_U lies on MA_U and E'_U lies on MA'_U, and therefore E_U and E'_U are different points.

(ii) All three points M, P_U, and P'_U lie on the same line (Fig. 3.14). Therefore, E_U and E'_U lie on this line, too. Then, we perform our assignment construction for both points P_U and P'_U as depicted in Fig. 3.12. We immediately see in Fig. 3.14 that E_U and E'_U are different.

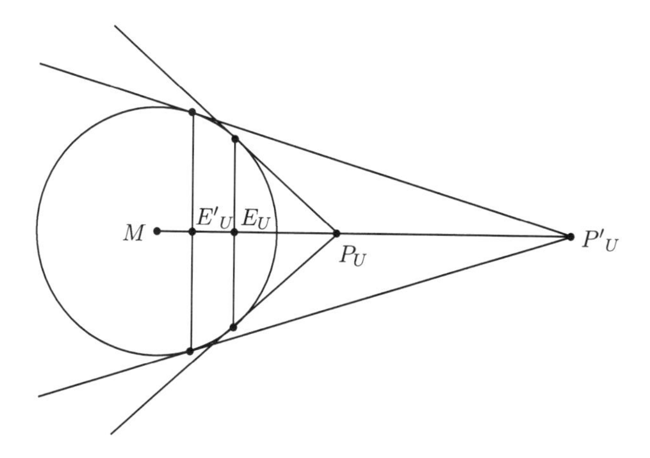

Fig. 3.14

We showed that, independently of the fact that the Universe is many times larger than Earth, the number of points in Earth is equal to the number of points in the Universe outside Earth. □

Exercise 3.4 Complete Fig. 3.13 by estimating the exact positions of points P_E and P'_E.

Exercise 3.5 Consider the semicircle in Fig. 3.15 and the line AB that is the diameter of the circle. Justify geometrically as well as by calculations that the number of points of the line AB is the same as the number of points of the curve of the semicircle.

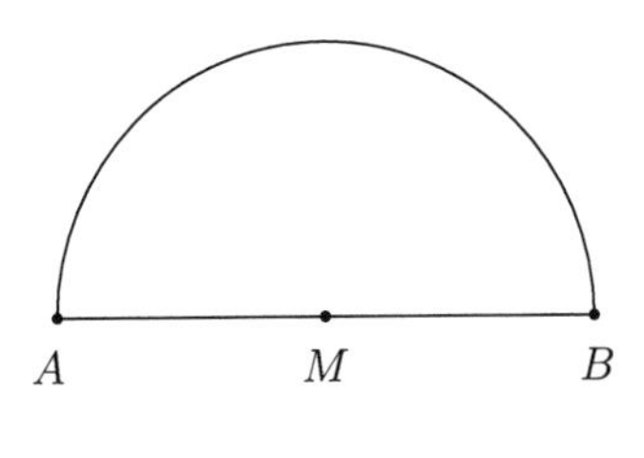

Fig. 3.15

Exercise 3.6 Consider the curve of the function F in Fig. 3.16 and the line AB. Why does this curve have as many points as the line AB?

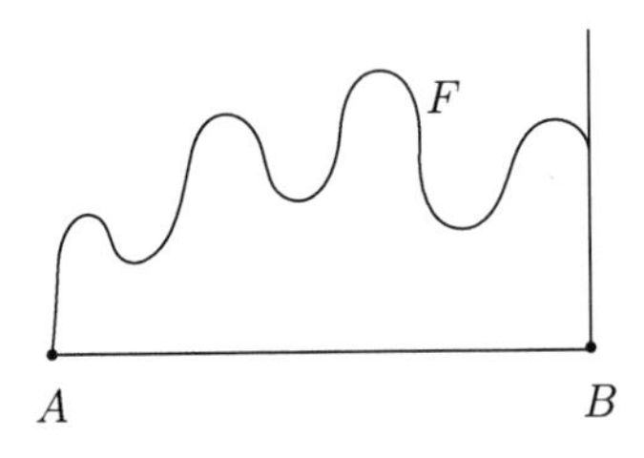

Fig. 3.16

If you still have a stomach ache when trying to imagine and to accept Cantor's concept of infinity, please, do not worry. Mathematicians needed many years to develop this concept, and, after discovering it, 20 years were needed to get it accepted by the broad mathematical community. Take time for repeated confrontations with the definition of infinite sets. Only if one iteratively deals with this topic can one understand why one uses this definition of infinity as an axiom of mathematics, and why mathematicians consider it not only trustworthy but they even do not see any alternative to this definition.

In what follows we briefly discuss the most frequent proposal for the concept of comparing infinite sizes that some listener proposed after the first confrontation with infinity. If

$$A \subset B$$

holds (i.e., if A is a proper subset of B), then

$$|A| < |B| \ .$$

Clearly, this attempt to compare infinite sizes reflects in another way the refusal of our key idea that a part of an infinite object may be as large as the whole. This proposal for an alternative definition has two drawbacks. First, one can use it only for comparing two sets where one is a subset of the other. This definition does not provide the possibility to compare two different sets such as $\mathbb{Z}^- = \{-1, -2, -3, \ldots\}$ and $\mathbb{Z}^+ = \{1, 2, 3, \ldots\}$. For a comparison of these two sets one has to search for another relation between them. Realizing this drawback, some listeners propose accepting the matching approach in the following way. One can find a matching between one of the sets and a subset of another one and then compare using the originally proposed subset principle. We show that one can get nonsense in this way. Namely that

$$|\mathbb{N}| < |\mathbb{N}| \ ,$$

i.e., that $\mathbb{N}$ is smaller than $\mathbb{N}$ itself. Using the concept of matching we proved

$$|\mathbb{N}| = |\mathbb{Z}^+|. \tag{3.1}$$

Since $\mathbb{Z}^+ \subset \mathbb{N}$, using the subset principle, one gets

$$|\mathbb{Z}^+| < |\mathbb{N}|. \tag{3.2}$$

Combining Equations (3.1) and (3.2) we obtain

$$|\mathbb{N}| = |\mathbb{Z}^+| < |\mathbb{N}|,$$

and so $|\mathbb{N}| < |\mathbb{N}|$.

In this way we proved that the concept of the shepherd (of matching) and the subset principle for comparing the cardinalities of two sets contradict each other because adopting both at once leads to obvious nonsense.

Why do we spend so much time discussing this axiom of mathematics and why do we make such a big effort to understand it? As you may already suspect, this axiom is only the beginning of our troubles. The concept of infinity is not the only surprise of this chapter. In some sense we showed $\infty = \infty + 1$ for $\infty = |\mathbb{N}|$ and also came to understand that $\infty = \infty + c$ for any finite number c. Example 3.2 and the following exercises even intimate

$$\infty = c \cdot \infty$$

for an arbitrary finite number (constant) c.

Let us consider $\mathbb{N}$ and the set

$$\mathbb{N}_{even} = \{0, 2, 4, 6, \ldots\} = \{2i \mid i \in \mathbb{N}\}$$

of all even natural numbers. At first glance $\mathbb{N}$ contains twice as many elements as $\mathbb{N}_{even}$. In spite of this view (Fig. 3.17) one can match the elements of $\mathbb{N}$ and of $\mathbb{N}_{even}$ as follows:

$$(0,0), (1,2), (2,4), (3,6), \ldots, (i, 2i), \ldots .$$

We see that each element of both sets is married exactly once. The immediate consequence is

$$|\mathbb{N}| = |\mathbb{N}_{even}| .$$

We can explain this somewhat surprising result

$$2 \cdot \infty = \infty$$

again using a story about Hotel Hilbert.

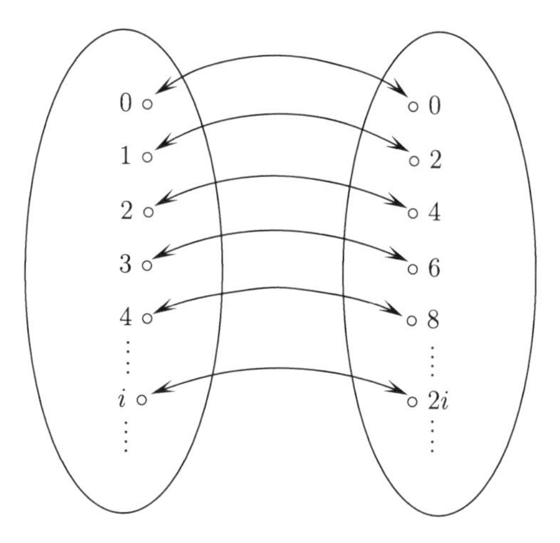

Fig. 3.17

Example 3.3 Consider once again Hotel Hilbert with infinitely many single rooms

$$Z(0), Z(1), Z(2), \ldots$$

that are all occupied by guests. Now, an infinite bus arrives. This bus has infinitely many seats

$$B(0), B(1), B(2), \ldots ,$$

and all seats are occupied by passengers[9]. The bus driver asks the porter whether he can accommodate all the passengers. As usual, the porter answers: "No problem", and does the following:

He asks each guest in room $Z(i)$ to move to room $Z(2i)$ as depicted in the upper part of Fig. 3.18. After the move, each former guest has her or his own room and all rooms with odd numbers $1, 3, 5, 7, \ldots, 2i + 1 \ldots$ are empty. Now, it remains to match the free rooms with the bus passengers. The porter assigns room $Z(1)$ to the passenger sitting on seat $B(0)$, room $Z(3)$ to the passenger sitting on sit $B(1)$, etc. In general, the passenger from $B(i)$ gets room $Z(2i + 1)$, as depicted in Fig. 3.18. In this way, one gets the matching

[9] Each seat is occupied by exactly one passenger.

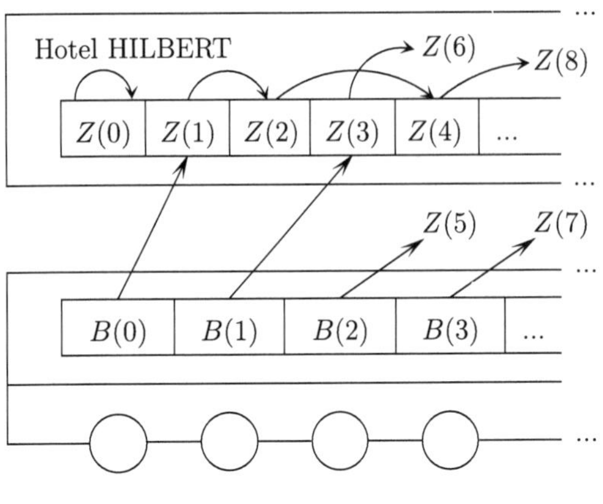

Fig. 3.18

$$(B(0), Z(1)), (B(1), Z(3)), (B(2), Z(5)), \ldots, (B(i), Z(2i+1)), \ldots$$

between the empty rooms with odd numbers and the seats of the infinite bus. □

Exercise 3.7 (a) Hotel Hilbert is only partially occupied. All rooms $Z(0), Z(2), Z(4), \ldots, Z(2i), \ldots$ with even numbers are occupied and all rooms with odd numbers are free. Now, two infinite buses B_1 and B_2 arrive. The seats of the buses are numbered as follows:

$$B_1(0), B_1(1), B_1(2), B_1(3), \ldots$$

$$B_2(0), B_2(1), B_2(2), B_2(3), \ldots$$

How can the porter act in order to accommodate all guests? Is it possible to accommodate all newcomers without asking somebody to move to another room?

(b) Hotel Hilbert is fully occupied. Now, three infinite buses are coming. The seats of each bus are enumerated by natural numbers. How can the porter accommodate everybody?

Exercise 3.8 Show by matching $\mathbb{Z}$ and $\mathbb{N}$ that

$$|\mathbb{Z}| = |\mathbb{N}|$$

holds, where $\mathbb{Z} = \{\ldots, -3, -2, -1, 0, 1, 2, 3, \ldots\}$ is the set of all integers.

Exercise 3.9 (challenge) Let $[a, b]$ be the set of all points (all real numbers) of the real axis between a and b.

a) Show that

$$|[0, 1]| = |[1, 10]| \ .$$

Try to show this by geometric means as in Example 3.2.

b) Prove

$$|[0,1]| = |[1,100]|$$

by arithmetic arguments, i.e., find a function f such that the pairs $(f(i), i)$ for $i \in [0, 100]$ build a matching of $[0, 1]$ and $[0, 100]$.

Exercise 3.10 (challenge) Assume that Hotel Hilbert is empty, i.e., there are no guests accommodated in the hotel. Since all used accommodation strategies were based on moving former guests from a room to another, there is the risk that to stay in the hotel may become unpopular. Therefore, the porter needs an accommodation strategy that does not require any move of an already accommodated guest. This accommodation strategy has to work even if arbitrarily many finite and infinite buses arrive in arbitrarily many different moments. Can you help the porter?

We observe that proving

$$|\mathbb{N}| = |A|$$

for a set A means nothing other than numbering all elements of set A using natural numbers. A matching between $\mathbb{N}$ and A unambiguously assigns a natural number from $\mathbb{N}$ to each element of A. And this assigned natural number can be viewed as the order of the corresponding element of A. For instance, if (3, John) is a pair of the matching, then John can be viewed as the third element of set A. Vice versa, each numbering of elements of a set A directly provides a matching between $\mathbb{N}$ and A. The pairs of the are simply

$$\text{order of } a, a$$

for each element a of A. In what follows, the notion of **numbering**[10] the elements of A enables us to present transparent arguments for claims $|\mathbb{N}| = |A|$ for some sets A, i.e., to show that A has as many elements as $\mathbb{N}$.

The matching

$$(0, 0), (1, 1), (2, -1), (3, 2), (4, -2), (5, 3), (6, -3), \ldots$$

of the sets $\mathbb{N}$ and $\mathbb{Z}$ assigns the following order to the elements of $\mathbb{Z}$:

$$0, 1, -1, 2, -2, 3, -3, \ldots \ .$$

In this way 0 is the 0-th element, 1 is the first element, -1 is the second element, 2 is the third element, etc.

[10] In the scientific literature one usually uses the term "enumeration" of the set A.

Exercise 3.11 Assign to $\mathbb{Z}$ an order of elements other than the one presented above by giving another matching.

Exercise 3.12 Prove that

$$|\mathbb{N}| = |\mathbb{N}_{quad}| ,$$

where $\mathbb{N}_{quad} = \{i^2 \mid i \in \mathbb{N}\} = \{0, 1, 4, 9, 16, 25, \ldots\}$ is the set of all squares of natural numbers. What order of the elements of $\mathbb{N}_{quad}$ do you get using the matching you proposed?

Our attempt to answer the next question increases the degree of difficulty of our considerations. What is the relation between $|\mathbb{N}|$ and $|\mathbb{Q}^+|$? Remember that

$$\mathbb{Q}^+ = \left\{ \frac{p}{q} \mid p, q \in \mathbb{Z}^+ \right\}$$

is the set of all positive rational numbers. We have already observed that by calculating averages repeatedly one can show that there are infinitely many rational numbers between any two rational numbers a and b with $a < b$. If one partitions the real axes into infinitely many parts $[0, 1], [1, 2], [2, 3], \ldots$ as depicted in Fig. 3.19, then the cardinality of $\mathbb{Q}^+$ looks like

$$\infty \cdot \infty = \infty^2$$

because each of these infinitely many parts (intervals) contains infinitely many rational numbers.

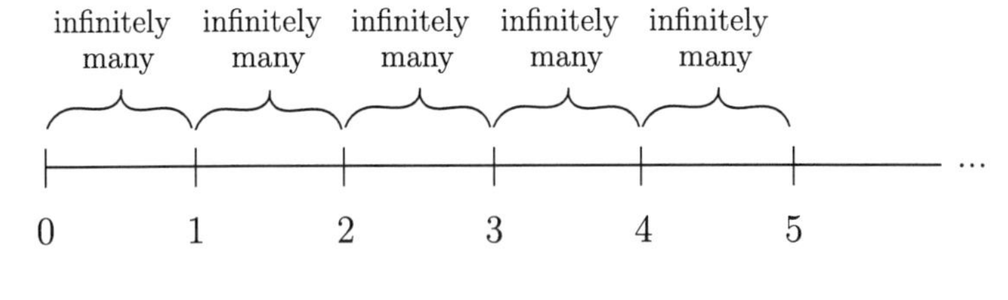

Fig. 3.19

At first glance, trying to prove the equality $|\mathbb{N}| = |\mathbb{Q}^+|$ does not seem very promising. The natural numbers $0, 1, 2, 3, \ldots$ lie very thinly on the right half of the axes, and between any two consecutive natural numbers i and $i + 1$ there are infinitely many rational numbers. Additionally, we know that a matching between $\mathbb{N}$ and $\mathbb{Q}^+$ would provide a numbering of elements in $\mathbb{Q}^+$. What does such a numbering of positive rational numbers look like? It cannot follow the size of the rational numbers, because, as we know, there is no smallest positive rational number[11].

Through this very clear impression, we show that the equality

$$|\mathbb{N}| = |\mathbb{Q}^+|$$

and so, in some sense, that

$$\infty \cdot \infty = \infty$$

holds.

Observe first that the set $\mathbb{Z} = \{\ldots, -3, -2, -1, 0, 1, 2, 3, \ldots\}$ also does not have any smallest number, and though we can number their elements as follows:

$$0, -1, 1, -2, 2, -3, 3, \ldots .$$

The idea for $\mathbb{Q}^+$ is to write all positive rational numbers on an infinite sheet as follows (the mathematicians among us would say that one assigns positions of the two-dimensional infinite matrix to positive rational numbers). Each positive rational number can be written as

$$\frac{p}{q},$$

where p and q are positive integers. We partition the infinite sheet of paper into infinitely many columns and infinitely many rows. We number the rows by

$$1, 2, 3, 4, 5, \ldots$$

from top to bottom, and we number the columns from left to right (Fig. 3.20). We place the fraction

$$\frac{i}{j}$$

[11] For any small rational number a, one can get the smaller rational number $a/2$ by halving a.

	1	2	3	4	5	6	...
1	$\frac{1}{1}$	$\frac{1}{2}$	$\frac{1}{3}$	$\frac{1}{4}$	$\frac{1}{5}$	$\frac{1}{6}$	...
2	$\frac{2}{1}$	$\frac{2}{2}$	$\frac{2}{3}$	$\frac{2}{4}$	$\frac{2}{5}$	$\frac{2}{6}$	...
3	$\frac{3}{1}$	$\frac{3}{2}$	$\frac{3}{3}$	$\frac{3}{4}$	$\frac{3}{5}$	$\frac{3}{6}$	...
4	$\frac{4}{1}$	$\frac{4}{2}$	$\frac{4}{3}$	$\frac{4}{4}$	$\frac{4}{5}$	$\frac{4}{6}$	...
5	$\frac{5}{1}$	$\frac{5}{2}$	$\frac{5}{3}$	$\frac{5}{4}$	$\frac{5}{5}$	$\frac{5}{6}$	...
6	$\frac{6}{1}$	$\frac{6}{2}$	$\frac{6}{3}$	$\frac{6}{4}$	$\frac{6}{5}$	$\frac{6}{6}$	...
$\vdots$	$\vdots$	$\vdots$	$\vdots$	$\vdots$	$\vdots$	$\vdots$	$\ddots$

Fig. 3.20

on the square in which the i-th row intersects the j-th column. In this way we get the infinite matrix as described in Fig. 3.20.

We do not have any doubt that this infinite sheet (this infinite matrix) contains all positive fractions. If one looks for an arbitrary fraction p/q, one immediately knows that p/q is placed on the intersection of the p-th row and the q-th column. But we have another problem. Some[12] positive rational numbers occur in the matrix several times, even infinitely many times. For instance, the number 1 can be represented as a fraction in the following different ways:

$$\frac{1}{1}, \frac{2}{2}, \frac{3}{3}, \frac{4}{4}, \ldots .$$

The rational number $1/2$ can be written as

$$\frac{1}{2}, \frac{2}{4}, \frac{3}{6}, \frac{4}{8}, \ldots .$$

Exercise 3.13 Which infinitely many representations as a fraction does the rational number $\frac{3}{7}$ have?

[12] in fact all

But we aim to have each positive rational number appearing exactly once on this sheet. Therefore, we take the fraction p/q that cannot be reduced[13] as a unique representation of the rational number p/q. In this way 1 uniquely represents $1/1$, one half is represented by $1/2$, because all other fractions represented by 1 and $1/2$ can be reduced. Hence, we remove (rub out) all fractions of the sheet that can be reduced. In this way we get empty positions (squares) on the intersections of some rows and columns, but this does not disturb us.

	1	2	3	4	5	6	...
1	$\frac{1}{1}$	$\frac{1}{2}$	$\frac{1}{3}$	$\frac{1}{4}$	$\frac{1}{5}$	$\frac{1}{6}$	...
2	$\frac{2}{1}$		$\frac{2}{3}$		$\frac{2}{5}$		...
3	$\frac{3}{1}$	$\frac{3}{2}$		$\frac{3}{4}$	$\frac{3}{5}$		...
4	$\frac{4}{1}$		$\frac{4}{3}$		$\frac{4}{5}$		...
5	$\frac{5}{1}$	$\frac{5}{2}$	$\frac{5}{3}$	$\frac{5}{4}$		$\frac{5}{6}$	...
6	$\frac{6}{1}$				$\frac{6}{5}$		...
$\vdots$	$\vdots$	$\vdots$	$\vdots$	$\vdots$	$\vdots$	$\vdots$	$\ddots$

Fig. 3.21

Now we want to number the fractions in Fig. 3.21 as the first, the second, the third, etc. Clearly, we cannot do it in the way in which first the elements (fractions) of the first row are numbered, then the elements of the second row, etc., since the number of elements in the first row is infinite. We would fail in such an attempt because we could never start to number the elements of the second row. The first row would simply consume all numbers of $\mathbb{N}$. Analogously, it is impossible to number the elements of the infinite sheet column by column. What can we do then? We number the elements of the

[13] The greatest common divisor of p and q is 1.

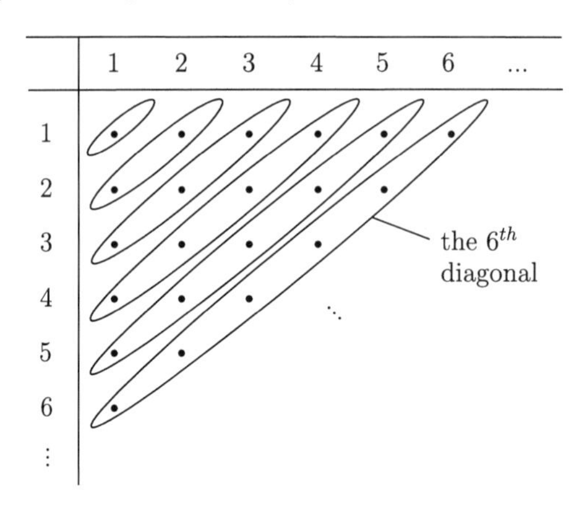

Fig. 3.22

sheet in Fig. 3.21 diagonal by diagonal. The ***k*-th diagonal of the sheet** contains all positions (Fig. 3.22) for which the sum of its row number i and its column number j is $k+1$ ($i+j=k+1$).

In this way the first diagonal contains only one element, $\frac{1}{1}$. The second diagonal contains two elements, $\frac{2}{1}$ and $\frac{1}{2}$. And, for instance, the fourth diagonal contains the four elements $\frac{4}{1}, \frac{3}{2}, \frac{2}{3}$, and $\frac{1}{4}$. In general, for each positive integer k, the k-th diagonal contains exactly k positions, and so at most k fractions.

Now, we order (number) the positions of the infinite sheet, and in this way we order the fractions lying there as shown in Fig. 3.23.

We order the diagonals according to their numbers, and we order the elements of any diagonal from the left to the right. Following this strategy and the placement of the fractions in Fig. 3.21, we obtain the following numbering of all positive rational numbers:

$$\frac{1}{1}, \frac{2}{1}, \frac{1}{2}, \frac{3}{1}, \frac{1}{3}, \frac{4}{1}, \frac{3}{2}, \frac{2}{3}, \frac{1}{4}, \frac{5}{1}, \frac{1}{5}, \frac{6}{1}, \frac{5}{2}, \frac{4}{3}, \frac{3}{4}, \frac{2}{5}, \frac{1}{6}, \ldots.$$

Following our numbering convention, $1/1$ is the 0-th rational number, $2/1$ is the first positive rational number, etc. For instance, $3/1$ is the third rational number, and $5/2$ is the 12-th one.

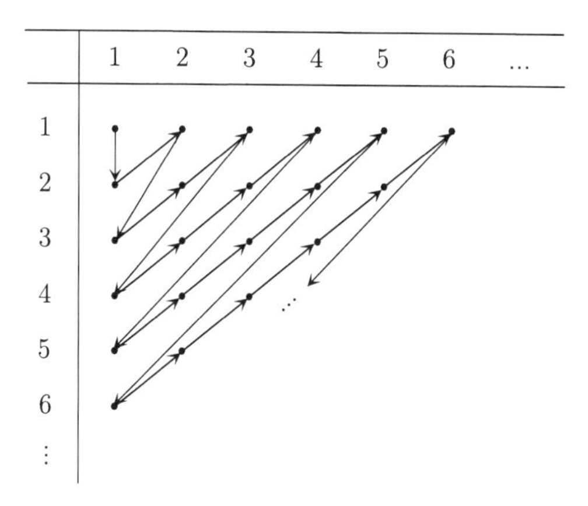

Fig. 3.23

Exercise 3.14 Extend the matrix in Fig. 3.21 by two more rows and columns and place the corresponding fractions in their visible positions. Use this extended matrix to write the sequence of fractions that got the orders $17, 18, 19, \ldots, 26, 27$ by our numbering.

The most important observation for seeing the correctness of our numbering strategy is that each positive rational number (fraction) is assigned a natural number as its order. The argument is straightforward. Let p/q be an arbitrary positive fraction. The rational number p/q is placed on the intersection of the p-th row and the q-th column, and so it lies on the diagonal $(p+q-1)$. Because **each diagonal contains finitely many positions (fractions)**, the numbering of elements of the forthcoming diagonals $1, 2, 3, \ldots, p+q-2$ is completed in a finite time, and so the numbering of the elements of the diagonal $p+q-1$ is performed too. In this way, p/q as an element of the diagonal $p+q-1$ is also given an order. Since the i-th diagonal contains at most i rational numbers, the order of p/q is at most

$$1+2+3+4+\ldots+(p+q-1) .$$

In this way, one can conclude that

$$|\mathbb{Q}^+| = |\mathbb{N}|$$

holds.

Exercise 3.15 Figure 3.24 shows another strategy for numbering of positive rational numbers that is also based on the consecutive numbering of diagonals. Write the first 20 positive rational numbers with respect to this numbering. What order is assigned to the fraction 7/3? What order does the number 7/3 have in our original numbering following the numbering strategy depicted in Fig. 3.23?

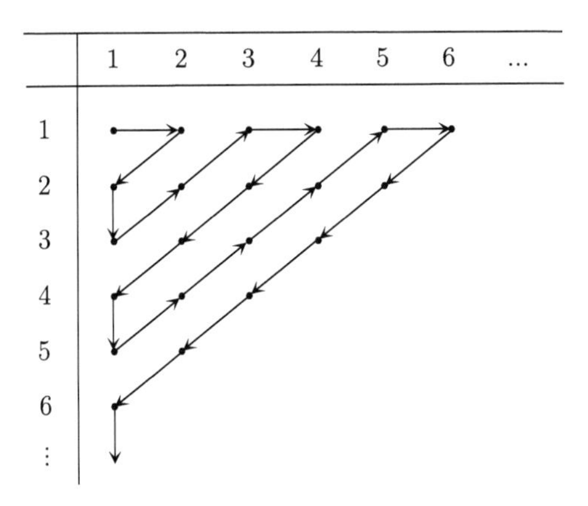

Fig. 3.24

Exercise 3.16 Hotel Hilbert is completely empty; that is no guest is staying there. At once (as sometimes happens in real life), infinitely many infinite buses arrive. The buses are numbered as

$$B_0, B_1, B_2, B_3, \dots ,$$

i.e., there are as many buses as $|\mathbb{N}|$. For each $i \in \mathbb{N}$, bus B_i contains infinitely many seats

$$B_i(0), B_i(1), B_i(2), B_i(3), \dots .$$

Each seat is occupied by exactly one passenger. How can the hotel porter accommodate all the passengers?

Exercise 3.17 (challenge) Prove that $|\mathbb{Q}| = |\mathbb{N}|$.

Exercise 3.18 (challenge) We define

$$\mathbb{N}^3 = \{(i, j, k) \mid i, j, k \in \mathbb{N}\}$$

as the set of all triples (i, j, k) of natural numbers. One can place any natural number on each of the three positions of a triple. Hence, one could say that $|\mathbb{N}^3| = |\mathbb{N}| \cdot |\mathbb{N}| \cdot |\mathbb{N}| = \infty \cdot \infty \cdot \infty = \infty^3$. Show that $|\mathbb{N}^3| = |\mathbb{N}|$, and so that $\infty = \infty^3$ holds.

3.3 There Are Different Infinite Sizes, or Why There Are More Real Numbers than Natural Ones

In Section 3.2 we learned Cantor's concept for comparing the cardinalities of sets. Surprisingly, we discovered that the property distinguishing infinite objects from finite ones is that infinite objects contain proper parts that are as large as the whole. We were unsuccessful in searching for an infinity that is larger than $|\mathbb{N}| = \infty$. Even the unexpected equality $|\mathbb{Q}^+| = |\mathbb{N}|$ holds. This is true even though the rational numbers are infinitely more densely placed on the real axis than the natural ones. This means that $\infty \cdot \infty = \infty$. For each positive integer i, one can even prove that the infinite number

$$\underbrace{|\mathbb{N}| \cdot |\mathbb{N}| \cdot \ldots \cdot |\mathbb{N}|}_{k\ times} = \underbrace{\infty \cdot \infty \cdot \ldots \cdot \infty}_{k\ times} = \infty^k$$

is again the same as $|\mathbb{N}| = \infty$.

We are not far from believing that all infinite sets are of the same size. The next surprise is that the contrary is true. In what follows we show that

$$|\mathbb{R}^+| > |\mathbb{N}| .$$

Before reading Section 3.2 one would probably believe that the number of real numbers is greater than the number of natural numbers. But now we know that $|\mathbb{Q}^+| = |\mathbb{N}|$ holds. And the real numbers have similar properties to those of the rational numbers. There is no smallest positive real number, and there are infinitely many real numbers on the real axis between any two different real numbers. Since $|\mathbb{N}| = |\mathbb{Q}^+|$, the inequality $|\mathbb{R}^+| > |\mathbb{N}|$ would directly imply

$$|\mathbb{R}^+| > |\mathbb{Q}^+| .$$

Is this not surprising? Later, in Chapter 4, we will get a deeper understanding of the difference between the sets $\mathbb{R}$ and $\mathbb{Q}$ that is also responsible for the truthfulness of $|\mathbb{R}| > |\mathbb{Q}|$. For now, we reveal only the idea that, in contrast to real numbers, all rational numbers have a finite representation as fractions. Most of the real numbers do not possess any finite description. In order to prove $|\mathbb{R}^+| > |\mathbb{N}|$, we prove a stronger result. Let $[0, 1]$ be the set of all real numbers between 0 and 1, the numbers 0 and 1 included. We show

$$|[0, 1]| \neq |\mathbb{N}| .$$

How can one prove inequality between the cardinalities (sizes) of two infinite sets? For proving equality, one has to find a matching between the two sets considered. This can be complicated, but in some sense it is easy because this is constructive. You find a matching and the work is done. To prove $|A| \neq |B|$ you have to prove that **there does not exist any matching between *A* and *B***. The problem is that there may exist infinitely many strategies for constructing a matching between A and B. How can you exclude the success of any of theses strategies? You cannot check all these infinitely many approaches one after another. When one has to show that something does not exist, then we speak about **proofs of nonexistence**.

> *To prove the nonexistence of an object or the impossibility of an event is the hardest task one can pose to a researcher in the natural sciences.*

The word "impossible" is almost forbidden in this context, and if one uses it then we have to be careful about its exact interpretation. A well-known physicist told me that it is possible to reconstruct the original egg from an egg fried in the pan. All is based on the reversibility of physical processes[14] and he was even able to calculate the probability of success for the attempt to create the original egg. The probability was so small that one could consider success as a real miracle, but it was greater than 0. There are many things considered impossible, though they are possible.

In mathematics we work in an artificial world; because of that we are able to create many proofs of nonexistence of mathematical objects. What remains is the fact that proofs of nonexistence belong to the hardest argumentations in mathematics.

[14] as formulated by quantum mechanics

Let us try to prove that it is impossible to number all real numbers from the interval $[0, 1]$, and so that $|[0, 1]| \neq |\mathbb{N}|$. As already mentioned, we do it by indirect argumentation. We assume that there is a numbering of real numbers from $[0, 1]$, and then we show that this assumption leads to a contradiction, i.e., that a consequence of this assumption is evident nonsense[15].

If there is a numbering of real numbers in $[0, 1]$ (i.e., if there is a matching between $[0, 1]$ and $\mathbb{N}$), then one can make a list of all real numbers from $[0, 1]$ in a table as shown in Fig. 3.25.

	0	1	2	3	4	$\dots$	i	$\dots$
1	0.	$\boxed{a_{11}}$	a_{12}	a_{13}	a_{14}	$\dots$	a_{1i}	$\dots$
2	0.	a_{21}	$\boxed{a_{22}}$	a_{23}	a_{24}	$\dots$	a_{2i}	$\dots$
3	0.	a_{31}	a_{32}	$\boxed{a_{33}}$	a_{34}	$\dots$	a_{3i}	$\dots$
4	0.	a_{41}	a_{42}	a_{43}	$\boxed{a_{44}}$	$\dots$	a_{4i}	$\dots$
$\vdots$	$\vdots$	$\vdots$	$\vdots$	$\vdots$	$\vdots$	$\ddots$		
i	0.	a_{i1}	a_{i2}	a_{i3}	a_{i4}	$\dots$	a_{ii}	$\dots$
$\vdots$	$\vdots$	$\vdots$	$\vdots$	$\vdots$	$\vdots$	$\vdots$	$\vdots$	$\ddots$

Fig. 3.25

This means that the first number in the list is

$$0.a_{11}a_{12}a_{13}a_{14} \dots .$$

The symbols $a_{11}, a_{12}, a_{13}, \dots$ are digits. In this representation, a_{11} is the first digit to the right of the decimal point, a_{12} is the second digit, a_{13} is the third one, etc. In general

$$0.a_{i1}a_{i2}a_{i3}a_{i4} \dots$$

[15] Here, we recommend revisiting the schema of indirect proofs presented in Chapter 1. If a consequence of an assertion Z is nonsense or contradicts something known, then the indirect proof schema says that Z does not hold, i.e., that the contrary of Z holds. The contrary of the existence of a matching between $[0, 1]$ and $\mathbb{N}$ is the nonexistence of any matching between $[0, 1]$ and $\mathbb{N}$.

is the i-th real number from $[0, 1]$ in our list (numbering). Our table is infinite in both directions. The number of rows is $|\mathbb{N}|$ and the number of columns is also $|\mathbb{N}|$, where the j-th column contains j-th digits behind the decimal points of all numbered real numbers in the list. The number of columns must be infinite, because most real numbers cannot be represented exactly by a bounded number of decimal positions behind the decimal point. For instance, the representation of the fraction

$$\frac{1}{3} = 0.\overline{3} = 0.33333\ldots$$

requires infinitely many digits to the right of the decimal point. On the other hand, this real number is nice because it is periodic. Numbers such as $\sqrt{2}/2$ and $\pi/4$ are not periodic and require infinitely many positions behind the decimal point for their decimal representation.

To be more transparent, we depict a concrete fraction of a hypothetical list of all real numbers from $[0, 1]$ in Fig. 3.26 by exchanging the abstract symbols a_{ij} for concrete digits.

	0	1	2	3	4	5	6	...
1	0.	[7]	3	2	1	1	0	...
2	0.	0	[0]	0	0	0	0	...
3	0.	9	9	[8]	1	0	3	...
4	0.	2	3	4	[0]	7	8	...
5	0.	3	5	0	1	[1]	2	...
6	0.	3	1	4	0	5	[7]	...
⋮	⋮	⋮	⋮	⋮	⋮	⋮	⋮	⋱
i	0.	7	6	5	0	0	1	...
⋮	⋮	⋮	⋮	⋮	⋮	⋮	⋮	⋱

Fig. 3.26

In this hypothetical list the number 0.732110... is the first real number, 0.000000... is the second real number, etc.

In what follows, we apply the so-called **diagonalization method** in order to show that there is a real number from $[0, 1]$ missing in the list (Fig. 3.25). This contradicts our assumption that one has a numbering of the elements of $[0, 1]$ (i.e., each number from $[0, 1]$ has to occur in the list exactly once). Hence, our hypothetical numbering is not a numbering, and we are allowed to conclude that there does not exist any numbering of the elements from $[0, 1]$.

Next, we construct a number c from $[0, 1]$ that is not represented by any row of the table (list), i.e., that differs from all numbers of the list. We create c digit by digit. We write c as

$$c = 0.c_1c_2c_3c_4 \ldots c_i \ldots \ ,$$

i.e., c_i is the i-th digit of c behind the decimal point. We choose $c_1 = a_{11} - 1$ if $a_{11} \neq 0$, and we set $c_1 = 1$ if $a_{11} = 0$. For the hypothetical numbering in Fig. 3.26 this means that $c_1 = 6$ because $a_{11} = 7$. Now we know with certainty that c is different from the number written in the first row of our list in Fig. 3.25 (Fig. 3.26). The second digit c_2 of c is again chosen in such a way that it differs from a_{22}. We take $c_2 = a_{22} - 1$ if $a_{22} \neq 0$, and we set $c_2 = 1$ if $a_{22} = 0$. Hence, c differs from the number in the second row of the list, and so c is not the second number of the hypothetical numbering. Next, one chooses c_3 in such a way that $c_3 \neq a_{33}$ in order to ensure that c is not represented by the third row of the list.

In general, one chooses $c_i = a_{ii} - 1$ for $a_{ii} \neq 0$, and $c_i = 1$ for $a_{11} = 0$. In this way c differs from the i-th number of our hypothetical numbering. After six construction steps for the table in Fig. 3.26 one gets

$$0.617106\ldots \ .$$

We immediately see that c differs from the numbers in the first 6 rows of the table in Fig. 3.26.

We observe that c differs from each number of the list in at least one decimal digit, and so c is not in the list. Therefore, the table in

Fig. 3.26 is not a numbering of $[0, 1]$. A numbering of $[0, 1]$ has to list all real numbers from $[0, 1]$, and c is clearly in $[0, 1]$. Hence, our assumption that one has a numbering of $[0, 1]$ (that there exists a numbering of $[0, 1]$) is false. We are allowed to conclude

There does not exist any numbering of $[0, 1]$, *and so there is no matching between* $\mathbb{N}$ *and* $[0, 1]$.

Exercise 3.19 Draw a table (as we did in Fig. 3.26) of a hypothetical numbering of $[0, 1]$ that starts with the numbers $1/4, 1/8, \sqrt{2}/2, 0, 1, \pi/4, 3/7$. Use this table to determine the digits $c_1, c_2, \ldots, c_7$ of the number c in such a way that c differs from the numbers in the first seven rows of your table.

Exercise 3.20 Consider a hypothetical numbering of $[0, 1]$, such that the 100-th number is $2/3$. Which digit of c is determined by this information?

Exercise 3.21 Determine the first seven digits of c behind the decimal point of a hypothetical numbering of $[0, 1]$ presented in Fig. 3.27.

	0	1	2	3	4	5	6	7	...
1	0.	[2]	0	0	1	7	8	0	...
2	0.	1	[7]	3	1	7	8	4	...
3	0.	1	6	[4]	3	3	3	3	...
4	0.	1	6	3	[0]	7	8	4	...
5	0.	1	6	3	1	[8]	8	4	...
6	0.	1	6	3	1	7	[9]	4	...
7	0.	1	6	3	1	7	8	[4]	...
⋮	⋮	⋮	⋮	⋮	⋮	⋮	⋮	⋮	⋱

Fig. 3.27

What exactly did we show and what was our argumentation? Assume somebody says, "I have a numbering of $[0, 1]$." We discovered a method, called diagonalization, that enables us to reject any proposal of a numbering of $[0, 1]$ as incomplete because at least one number from $[0, 1]$ is missing there. Since we can do it for each hypothetical numbering of the elements of $[0,1]$, there does not exist any (complete) numbering of $[0,1]$.

Another point of view is that of indirect argumentation introduced in Chapter 1. Our aim was to prove the claim Z that there does not exist any numbering of $[0,1]$. We start with the opposite claim $\overline{Z}$ and show that a consequence of $\overline{Z}$ is a claim which is nonsense. In this moment we have reached our goal. The assertion $\overline{Z}$ as the opposite of Z is the claim that there exists a numbering of the elements of $[0,1]$. Starting from $\overline{Z}$ we show that in any such numbering of $[0,1]$ one number from $[0,1]$ is missing. This is nonsense because no number is allowed to be missing in a numbering. Therefore, $\overline{Z}$ does not hold, and so there does not exist any numbering of $[0,1]$.

Since we cannot number the elements of $[0,1]$ (there is no matching between $\mathbb{N}$ and $[0,1]$), we cannot number the elements of $\mathbb{R}^+$ either.

Exercise 3.22 Explain why the nonexistence of a numbering of the elements of $[0,1]$ implies the nonexistence of a numbering of the elements of $\mathbb{R}^+$.

Hint: You can try to explain how to transform each numbering of $\mathbb{R}^+$ into a numbering of $[0,1]$. Why is this a correct argument?

Since $\mathbb{N} \subset \mathbb{R}^+$ and there is no matching between $\mathbb{N}$ and $\mathbb{R}^+$, we can conclude that

$$|\mathbb{N}| < |\mathbb{R}^+|$$

holds. Hence, there are at least two infinite sets of different sizes, namely $\mathbb{N}$ and $\mathbb{R}^+$. One can even show that there are unboundedly many (infinitely many) different infinite sizes. We do not to deal with the technical proof of this result here because we do not need it for reaching our main goal. We are ready to show in the next chapter that the number of computing tasks is larger than the number of algorithms, and so that there exist problems that cannot be solved algorithmically (automatically by means of computers).

Exercise 3.23 Let us change the diagonalization method presented in Fig. 3.25 a little bit. For each $i \in \mathbb{N}$, we choose $c_i = a_{i,2i} - 1$ for $a_{i,2i} \neq 0$ and $c_i = 1$ for $a_{i,2i} = 0$.

a) Are we allowed again to say that the number $0.c_1c_2c_3c_4\dots$ is not included in the list? Argue for your answer!
b) Frame the digits $a_{i,2i}$ of the table in Fig. 3.25.
c) Which values are assigned to c_1, c_2, and c_3 for the hypothetic list in Fig. 3.27 in this way? Explain why the created number $c = 0.c_1c_2c_3\dots$ is not among the first three numbers of the table.

3.4 The Most Important Ideas Once Again

Two infinite sizes can be compared. One has to represent them using the cardinalities of the two sets. Using this as a basis, Cantor introduced the concept for comparing infinite sizes (cardinalities) of two sets using the shepherd's principle. Two sets are equally sized if one can match their elements. A set A has the same cardinality as $\mathbb{N}$ if one can number all elements of A using natural numbers. Clearly, each numbering of A corresponds to a matching between A and $\mathbb{N}$. Surprisingly, one can match $\mathbb{N}$ and $\mathbb{Z}$, though $\mathbb{N}$ is a proper part of $\mathbb{Z}$. In this way we recognized that the property

having a proper part that is as large as the whole

is exactly the characteristic that enables us to distinguish finite objects from infinite ones. No finite object may have this property. For infinite objects, this is a must. Though there are infinitely many rational numbers between any two consecutive natural numbers i and $i+1$, we found a clever enumeration[16] of all positive rational numbers, and so we showed that $|\mathbb{N}| = |\mathbb{Q}^+|$. After that, we applied the schema of indirect proofs in order to show that there is no numbering of all positive real numbers, and so that there is no matching between $\mathbb{N}$ and $\mathbb{R}^+$.

In Chapter 4, it remains to show that the number of programs is equal to $|\mathbb{N}|$, and that the number of algorithmic tasks is at least $|\mathbb{R}^+|$.

In Chapter 3, we did not present any miracle of computer science. But we did investigate the nature of infinity and the concept of

[16] not according to their sizes

comparing infinite sizes, and in this way we learned miracles of mathematics that are real jewels of the fundamentals of science. Jewels are not found lying on the street, and one usually has to do something to obtain them. Therefore, we are also required to sweat a bit in order to grasp infinity. And so, one may not be surprised that taking our path to the computer science miracles can be strenuous. But tenacity is a good property and the goal is worth the effort. Let us stay this course in the next two chapters, and then we will witness one miracle after the other. We will experience unexpected and elegant solutions to hopeless situations that increase the pulse of each friend of science. Only by patience and hard work can one attain knowledge that is really valuable.

Solutions to Some Exercises

Exercise 3.1 For the sets $A = \{2, 3, 4, 5\}$ and $B = \{2, 5, 7, 11\}$ there are $4! = 24$ different matchings. For instance,

$$(2, 11), (3, 2), (4, 5), (5, 7)$$

or

$$(2, 11), (3, 7), (4, 5), (5, 2) \ .$$

The sequence of pairs $(2, 2), (4, 5), (5, 11), (2, 7)$ is not a matching between A and B because element 2 of A occurs in two pairs, $(2, 2)$ and $(2, 7)$, and element 3 of A does not occur in any pair.

Exercise 3.8 A matching between $\mathbb{N}$ and $\mathbb{Z}$ can be found in such a way that one orders the elements of $\mathbb{Z}$ in the following sequence

$$0, 1, -1, 2, -2, 3, -3, 4, -4, \ldots, i, -i, \ldots$$

and then creates a matching by assigning to each element of $\mathbb{Z}$ its order in this sequence. In this way we get the matching

$$(0, 0), (1, 1), (2, -1), (3, 2), (4, -2), (5, 3), (6, -3), \ldots \ .$$

In general we build the pairs

$$(0, 0), (2i, -i) \text{ and } (2i - 1, i)$$

for all positive integers i.

Exercise 3.10 (challenge) First, the porter partitions all rooms into infinitely many groups, each of an infinite size. Always when a group of guests arrives (it does not matter whether the group is finite or infinite), the porter accommodates the guest in the next (still unused) group of rooms.

As usual for the staff of Hotel Hilbert, the porter is well educated in mathematics, and so he knows that there are infinitely many primes

$$2, 3, 5, 7, 11, 13, 17, 19, \ldots .$$

Let p_i be the i-th prime of this sequence. The porter uses p_i to determine the i-th infinite group of natural numbers as follows:

$$\text{group}(i) = \{p_i, p_i^2, p_i^3, p_i^4, \ldots, (p_i)^j, \ldots\} .$$

For instance, group(2)= $\{3, 9, 27, 81, \ldots\}$. Due to his knowledge of the Fundamental Theorem of Arithmetics, the porter knows that no natural number belongs to more than one group. Using this partition of rooms into the groups with respect to their room numbers, the porter can assign the rooms to the guests without having any more rooms even when infinitely many groups of guests arrive one after each other. It does not matter whether the i-th group of guest is finite or infinite, the porter books the whole room group(i) for the i-th guest group. If the guests of the i-th group are denoted as

$$G_{i,1}, G_{i,2}, G_{i,3}, \ldots, G_{i,j}, \ldots$$

then guest $G_{i,1}$ gets the room $Z(p_i)$, guest $G_{i,2}$ gets room $Z(p_i^2)$, etc.

Exercise 3.12 The sequence of pairs

$$(0, 0), (1, 1), (2, 4), (3, 9), (4, 16), \ldots, (i, i^2), \ldots$$

is a matching between $\mathbb{N}$ and $\mathbb{N}_{quad}$. We see that each number from $\mathbb{N}$ appears exactly once as the first element in a pair, and analogously each integer from $\mathbb{N}_{quad}$ can be found exactly once as the second element of a pair.

Exercise 3.20 The decimal representation of the fraction 2/3 is

$$0.\overline{6} = 0.666666\ldots .$$

Hence, the 100-th position behind the decimal point is also 6. Therefore, one sets $c_{100} = 6 - 1 = 5$.

Exercise 3.21 For the hypothetical numbering of real numbers from interval $[0, 1]$ in Fig. 3.27, one gets

$$c = 0.1631783\ldots .$$

Exercise 3.22 We perform an indirect proof by following the schema of the indirect argumentation from Chapter 1. We know that there is no numbering of $[0, 1]$. The aim is to show that there does not exist any numbering of $\mathbb{R}^+$. Assume the contrary of our aim, i.e., that there is a numbering of $\mathbb{R}^+$. We consider this numbering of $\mathbb{R}^+$ as a list and erase those numbers of this list that are not from $[0, 1]$. What remains is the list of numbers from $[0, 1]$ that is (without any doubt) a numbering of $[0, 1]$. But we know that there does not exist any numbering of $[0, 1]$, and so the contrary of our assumption must hold. The contrary of our assumption is our aim, i.e., that there does not exist any numbering of $\mathbb{R}^+$.

Discuss, commit errors, make mistakes,
but for God's sake think –
even if you should be wrong –
but think your own thoughts.

Gotthold Ephraim Lessing

Chapter 4

Limits of Computability or Why Do There Exist Tasks That Cannot Be Solved Automatically by Computers

4.1 Aim

In Chapter 3 we discovered that there exist different infinite sizes. For instance, the number of real numbers is a larger infinity than the number of natural numbers. An infinite set is exactly as large as $\mathbb{N}$ if one can number the elements of A as the first one, the second one, the third one, etc. Here we aim to show that computing tasks exist that cannot be solved using any algorithm. The idea of our argument is simple. We show that the number of different tasks (computing problems) is a larger infinity than the number

J. Hromkovič, *Algorithmic Adventures*, DOI 10.1007/978-3-540-85986-4_4,

of all programs. Hence, there exist problems that cannot be algorithmically solved, and so their solution cannot be automatically found by means of computers. But it is not satisfactory to prove the existence of algorithmically unsolvable problems. One could think that all algorithmically unsolvable problems are so artificial that none of them is really interesting for us. Therefore, we strive to show that there are concrete problems of serious interest in practice that cannot be algorithmically solved.

This chapter is the hardest one of this book, and so do not worry or be frustrated when you do not get a complete understanding of all details. Many graduate students at universities do not master this topic in detail. It is already valuable if one is able to understand and correctly interpret the computer science discoveries presented in what follows. To gain full understanding of the way in which these results were discovered usually requires multiple readings and discussions of the proof ideas. How many confrontations with this hard topic you perform is up to you.

It is important to know that one can successfully study the topics of all following chapters even if you do not understand all the arguments of Chapter 4.

4.2 How Many Programs Exist?

How many programs do we have? The first simple answer is "Infinitely many." Clearly, for each program A, there is another program B that is longer by a row (by an instruction) than A. Hence, there are infinitely many program lengths and so infinitely many programs must exist. Our main question is whether the number of programs is equal to $|\mathbb{N}|$ or not. First we aim to show that the number of different programs is the same infinite size as the number of natural numbers. We show it by giving a number of programs.

Let us start by thinking about the number of texts that can be written using a computer or a typewriter. Each text can be viewed as a sequence of **symbols** of the keyboard used. We have to take into account all uppercase and lowercase letters of the Latin alphabet. Additionally, one is allowed to use symbols such as

?, !, ·, $, /, +, *, etc.

Moreover, every keyboard contains a key for the character blank. For instance, we use a blank to separate two words or two sentences. We often use the symbol ␣ to indicate the occurrence of the character blank. Since blank has its meaning in texts, we consider it as a symbol (letter). From this point of view, texts are not only words such as

"computer" or "mother"

and not only sentences such as

"Computer␣science␣is␣full␣of␣magic",

but also sequences of keyboard characters without any meaning such as

xyz*-+?!abe/ .

This means that we do not expect any meaning of a text. Semantics does not play any role in our definition of the notion of a text. A text is simply a sequence of symbols that does not need to have any interpretation. In computer science the set of symbols used is called an **alphabet**, and we speak about **texts over an alphabet** if all texts considered consist of symbols of this alphabet only.

Because blank is considered as a symbol, the content of any book may be viewed as a text. Hence, we can fix the following:

Every text is finite, but there is no upper bound on the length of a text. Therefore, there are infinitely many texts.

Let us observe the similarity to natural numbers. Each natural number has a finite decimal representation as a sequence of digits (symbols) $0, 1, 2, 3, 4, 5, 6, 7, 8,$ and 9. The length of the decimal representation grows with the number represented, and so there is no bound on the representation length of natural numbers[1]. Hence, natural numbers can be viewed as texts over the alphabet of decimal digits and so the number of texts over the alphabet $\{0, 1, 2, 3, 4, 5, 6, 7, 8, 9\}$ is equal to $|\mathbb{N}|$. If one uses the binary alphabet $\{0, 1\}$ to represent natural numbers, one sees that the number of texts over the binary alphabet is also equal to $|\mathbb{N}|$.

It appears that the size of the alphabet does not matter, and so one can conjecture that

> *"The number of all texts over the characters of a keyboard is equal to $|\mathbb{N}|$."*

This is true, and we prove it by enumerating the texts. It is sufficient to show that one can order all texts in an infinite list. The ordering works in a similar way to creating a dictionary, but not exactly in the same way. Following the sorting rules of the dictionary, we would have to take first the texts $a, aa, aaa, aaaa$, etc., and we would never order texts containing a symbol different from a, because there are infinitely many texts consisting of the letter a only. Therefore, we have to change the sorting approach a little bit. To order all texts in a list, we first apply the following rule:

> *Shorter texts are always before longer texts.*

This means that our infinite list of all texts starts with all texts of length 1, then texts of length 2 follow, after that texts of length 3, etc. What still remains is to fix the order of the texts of the same length for any length. If one uses the letters of the Latin alphabet only, then one can do it in the same way as is used in dictionaries. This means to start with texts that begin with the letter a, etc. Since we also have many special symbols on our keyboard such as ?, !, *, +, etc., we have to order the symbols of our keyboard alphabet first. Which order of symbols we use is our choice, and for our claim about the cardinality of the set of all texts over the keyboard alphabet it does not matter. Figure 4.1 depicts a possible

[1] This means that one cannot restrict the representation length by a concrete number. For instance, if one upperbounds the representation length by n, then one would have at most 10^n different representations available, and this is not enough to represent all infinitely many natural numbers

ordering of the symbols of the keyboard alphabet. Having an order of the alphabet symbols,

One sorts the texts of the same length in the same way as in dictionaries[2].

1	2	3	...	25	26	27	28	...	51	52	53	54	...	61	62
a	b	c	...	y	z	A	B	...	Y	Z	1	2	...	9	0

63	64	65	66	67	68	69	70	71	72	73	74	75	...	167
+	"	*	ñ	&	!	.	:	,	;	?	$	Ý	...	␣

Fig. 4.1

This means that for the same text length we start with texts beginning with the first symbol of our alphabet ordering. For instance, taking the order of symbols depicted in Fig. 4.1, the numbering of texts starts with

1 a
2 b
3 c
⋮
167 ␣
⋮

Then, the texts of length 5 are ordered as follows:

aaaaa
aaaab
aaaac
⋮
aaaa␣
aaaba
aaabb
aaabc
⋮

[2] Usually, we speak about the lexicographical order.

Why did we take the time to show that the number of texts is equal to $|\mathbb{N}|$? Because

Each program is a text over the keyboard alphabet.

Programs are nothing more than special texts that are understandable for computers. Therefore, the number of all programs is not larger than the number of all texts over the keyboard alphabet, and so we can claim

The number of programs is equal to $|\mathbb{N}|$.

What we really have shown is that the number of programs is infinite and not larger than $|\mathbb{N}|$. The equality between $|\mathbb{N}|$ and the number of programs is the consequence of the fact that $|\mathbb{N}|$ is the smallest infinite size. But we did not prove this fact. Hence, if we want to convince the reader and provide a full argumentation of this fact, then we have to find a matching between $\mathbb{N}$ and programs. As we already know, any numbering provides a matching. And

One gets a numbering of all programs by erasing all texts that do not represent any program from our infinite list of all texts over the keyboard alphabet.

It is important to observe that the deletion of texts without an interpretation as programs can even be done automatically. One can write programs, called **compilers**, that get texts as inputs and decide whether a given text is a program in the programming language considered or not. It is worth noting that

A compiler can check the syntactical correctness of a text as a program but not the semantical correctness.

This means that a compiler checks whether a text is a correctly written sequence of computer instructions, i.e., whether the text is a program. A compiler does not verify whether the program is an algorithm, i.e., whether the program does something reasonable or not, or whether the program can repeat a loop infinitely many times.

Hence, we are allowed to number all programs, and so to list all programs in a sequence

$$P_0, P_1, P_2, P_3, \ldots, P_i, \ldots$$

where $\boldsymbol{P_i}$ denotes the i**-th program**.

Why is it important for us that the number of programs and so the number of algorithms is not larger than $|\mathbb{N}|$? The answer is that the number of all possible computing tasks is larger than $|\mathbb{N}|$, and so there are more problems than algorithms. The immediate consequence is that there exist problems that cannot be solved by any algorithm (for which no solution method exists).

We have already shown in Chapter 3 that the number of problems is very large. For each real number c, one can consider the following computer task Problem(c).

Problem(c)

Input: a natural number n

Output: a number c up to n decimal digits after the decimal point

We say that an **algorithm A_c solves Problem(c)** or that $\boldsymbol{A_c}$ **generates c**, if, for any given $n \in \mathbb{N}$, A_c outputs all digits of c before the decimal point and the first n digits of c after the decimal point.

For instance,

- For $c = \frac{4}{3}$ and an input $n = 5$, the algorithm $A_{\frac{4}{3}}$ has to give the output 1.33333.
- For $\sqrt{2}$, the algorithm $A_{\sqrt{2}}$ has to generate the number 1.4142 for the input $n = 4$ and the number 1.414213 for the input $n = 6$.
- For $c = \pi$, the algorithm A_π has to provide the output 3.141592 for $n = 6$.

Exercise 4.1 What is the output of an algorithm $A_{\frac{17}{6}}$ generating 17/6 for the input $n = 12$? What are the outputs of an algorithm A_π that generates π for inputs $n = 2, n = 0, n = 7$, and $n = 9$?

Exercise 4.2 (challenge) Can you present a method for the generation of π up to an arbitrarily large number of digits after the decimal point?

In Chapter 3, we proved that the number of real numbers is larger than $|\mathbb{N}|$, i.e., that $|\mathbb{R}| > |\mathbb{N}|$. Since the number of algorithms is not larger than $|\mathbb{N}|$, the number of real numbers is larger than the number of algorithms. Therefore, we can conclude that

There exists a real number c such that Problem(c) is not algorithmically solvable.

Thus, we proved that there are real numbers that cannot be generated by any algorithm. Do we understand exactly what this means? Let us try to build our intuition in order to get a better understanding of this result. Objects such as natural numbers, rational numbers, texts, programs, recipes, and algorithms have something in common.

All these objects have a finite representation.

But this is not true for real numbers. If one can represent a real number in a finite way, then one can view this representation as a text. Since the number of different texts is smaller than the number of real numbers, there must exist a real number without any finite representation.

What does it mean exactly? To have a constructive description of a real number e means that one is able to generate e completely digit by digit. Also, if the number e has an infinite decimal representation, one can use the description to unambiguously estimate the digit on any position of its decimal representation. In this sense, the finite description of e is complete. In other words, such a finite description of e provides an algorithm for generating e. For instance, $\sqrt{2}$ is a finite description of the irrational number $e = \sqrt{2}$, and we can compute this number with an arbitrarily high precision using an algorithm.[3] Therefore, we are allowed to say:

[3] For instance, using the algorithm of Heron.

Real numbers having a finite representation are exactly the numbers that can be algorithmically generated, and there exist real numbers that do not possess a finite representation and so are not computable (algorithmically generable).

Exercise 4.3 What do you mean? Are there more real numbers with finite representations than real numbers without any finite representation, or vice versa? Justify your answer!

We see that there are tasks that cannot be solved by algorithms. But we are not satisfied with this knowledge. Who is interested in asking for an algorithm generating a number e that does not have any finite representation? How can one formulate such a task in a finite way? Moreover, when only tasks of this kind are not algorithmically solvable, then we are happy and forget about this "artificial" theory and dedicate our time to solving problems of practical relevance. Hence, you may see the reason why we do not stop our investigation here and are not contented with our achievements. We have to continue our study in order to discover whether there are interesting computing tasks with a finite description that cannot be automatically solved by means of computers.

4.3 YES or NO, That Is the Question, or Another Application of Diagonalization

Probably the simplest problems considered in computer science are decision problems. A decision problem is to recognize whether a given object has a special property we are searching for or not. For instance, one gets a digital picture and has to decide whether a chair is in the picture. One can also ask whether a person is in the picture, or even whether a specific person (for instance, Albert Einstein) is in the picture. The answer has to be an unambiguous "YES" or "NO". Other answers are not allowed and we impose the requirement that the answer is always correct.

Here we consider a simple kind of decision problems. Let M be an arbitrary subset of $\mathbb{N}$, i.e., let M be a set that contains some natural numbers. We specify the **decision problem** $(\mathbb{N}, M)$as follows.

Input: a natural number n from $\mathbb{N}$

Output:

"YES" if n belongs to M
"NO" if n does not belong to M

For instance, one can take PRIME as M, where

$$\text{PRIME} = \{2, 3, 5, 7, 11, 13, 17, 19, \ldots\}$$

is the infinite set of all primes. Then, ($\mathbb{N}$, PRIME) is the problem to decide whether a given natural number n is prime or not. The problem $(\mathbb{N}, \mathbb{N}_{even})$ is to decide whether a given nonnegative integer is even or not.

For each subset M of $\mathbb{N}$ we say that an **algorithm A recognizes M** or that an **algorithm A solves the decision problem $(\mathbb{N}, M)$**if, for any input n, A computes

(i) the answer "YES" if n belongs to M, and

(ii) the answer "NO" if n does not belong to M ($n \notin M$).

Sometimes one uses the digit "1" instead of "YES" and the digit "0" instead of "NO". If A answers "YES" for an input n, then we say that **algorithm A accepts the number n**. If A outputs "NO" for an input n, then we say that **algorithm A rejects the number n**. Thus, an algorithm recognizing PRIME accepts each prime and rejects each composite number.

If there exists an algorithm solving a decision problem $(\mathbb{N}, M)$, then we say that the problem $(\mathbb{N}, M)$ is **algorithmically solvable** or that

the problem $(\mathbb{N}, M)$ is decidable.

Clearly, the decision problem $(\mathbb{N}, \mathbb{N}_{even})$ is decidable. It is sufficient to verify whether a given natural number is even or odd. The problem ($\mathbb{N}$, PRIME) is also decidable because we know how to check whether a natural number is a prime or not, and it is not too complicated to describe such a method as an algorithm.

Exercise 4.4 The naive method for primality testing is to divide the given number n by all numbers between 2 and $n-1$. If none of these $n-2$ numbers divides n, then n is a prime. To test primality in this way means to perform a lot of work. For the number 1000002 one has to execute 1000000 divisibility tests. Can you propose another method that can verify primality by performing an essentially smaller number of divisibility tests?

Exercise 4.5 (challenge) Write a program in the programming language TRANSPARENT of Chapter 2 that solves the problem ($\mathbb{N}$, QUAD) where

$$\text{QUAD} = \{1, 4, 9, 16, 25, \ldots\}$$

is the set of all squares i^2.

First, we aim to show the existence of decision problems that are not algorithmically solvable. Such decision problems are called

undecidable or **algorithmically unsolvable**.

We already recognized that we can list all programs as $P_0, P_1, P_2, \ldots$ and later we will see that one can do it using an algorithm. To list all algorithms using an algorithm is not so easy. Therefore, we begin our effort by proving a stronger result than we really need. We show that there are decision problems that cannot be solved by any program. What does "solved by a program" mean? What is the difference between algorithmic solvability and solvability by a program?

Remember that each algorithm can be written as a program, but it does not hold that each program is an algorithm. A program can perform a pointless work. A program can perform infinite work for some inputs without producing any result. But an algorithm must always finish its work in a *finite time* and produce a *correct result.*

Let M be a subset of $\mathbb{N}$. We say that a **program P accepts the set P**, if, for any given natural number n,

(i) P outputs "YES", if n belongs to M, and

(ii) P outputs "NO" or works **infinitely long** if n does not belong to M.

For a program P, $\boldsymbol{M(P)}$ denotes the set M accepted by P. In this way, P can be viewed as a finite representation of the potentially infinite set $M(P)$.

Immediately, we see the difference between the recognition of M by an algorithm and the acceptance of M by a program. For inputs from M both the algorithm and the program are required to work correctly and provide the right answer "YES" in a finite time (see requirement (i)). In contrast to an algorithm, for numbers not belonging to M, a program is allowed to work infinitely long and so never produce any answer. In this sense algorithms are special programs that never run infinite computations. Therefore, it is sufficient to show that there is no program accepting a set M, and the direct consequence is that there does not exist any algorithm that recognizes M (i.e., solves the decision problem $(\mathbb{N}, M)$).

To construct such a "hard" subset M of $\mathbb{N}$, we use the diagonalization method from Chapter 3 again. For this purpose, we need the following infinite representation of subsets of natural numbers (Fig. 4.2).

	0	1	2	3	4	...	i	$i+1$	...
M	0	1	0	0	1	...	1	0	...

Fig. 4.2

M is represented as an infinite sequence of bits. The sequence starts with the position 0 and has 1 at the i-th position if and only if the number i is in M. If i is not in M, then the bit 0 is on the i-th position of the sequence. The set M in Fig. 4.2 contains the numbers $1, 4$, and i. The numbers $0, 2, 3$, and $i+1$ are not in M. The binary representation of $\mathbb{N}_{even}$ looks as follows

10101010101010101010 ...

The representation of PRIM starts with the following bit sequence:

0011010100010100 ...

Exercise 4.6 Write out the first 17 bits of the binary representation of QUAD.

Now, we again build a two-dimensional table that is infinite in both directions. The columns are given by the infinite sequence of all numbers:

$$0, 1, 2, 3, 4, 5, \ldots, i, \ldots.$$

The rows are given by the infinite sequence of all programs:

$$P_0, P_1, P_2, P_3, \ldots, P_i, \ldots$$

that reads an input number only once and their only possible outputs are "YES" and "NO". One can recognize such programs by determining whether they contain only one instruction "to rest" and whether the only output instructions are writing the text "YES" or the text "NO". Each such program P_i unambiguously defines a set $M(P_i)$ of all natural numbers that are accepted[4] by P_i. Those numbers, for which P_i outputs "NO" or works infinitely long do not belong to $M(P_i)$.

The rows of our table are the binary representations of sets $M(P_i)$. The k-th row (see Fig. 4.3) contains the binary representation of the set $M(P_k)$ that is accepted by the program P_k. The intersection of the i-th row and the j-th column contains "1" if P_i accepts the number j (if P_i halts on the input j with the output "YES"). The symbol "0" lies in the intersection of the i-th row and the j-th column, if P_i outputs "NO" or works infinitely long for the input j. Hence

The infinite table contains in its rows the representation of ***all*** *subsets of* $\mathbb{N}$ *that can be accepted by a program.*

Next we aim to show that there is at least one subset of $\mathbb{N}$ missing in the table, i.e., that there is a subset of $\mathbb{N}$ whose binary

[4] P_i finishes the work on them by printing "YES".

	0	1	2	3	4	5	6	...	i	...	j	...
$M(P_0)$	0	1	1	0	0	1	0		1		0	
$M(P_1)$	0	1	0	0	0	1	1		0		0	
$M(P_2)$	1	1	1	0	0	1	0		1		1	
$M(P_3)$	1	0	1	0	1	0	1		1		0	
$M(P_4)$	0	0	0	1	1	0	1		0		1	
$M(P_5)$	1	1	1	1	1	1	1		1		1	
$M(P_6)$	1	0	1	0	0	0	1		0		1	
$\vdots$												...
$M(P_i)$	0	1	1	0	0	1	0		1			
$\vdots$												...
$M(P_j)$	1	0	1	0	1	1	1				0	
$\vdots$												$\vdots$

Fig. 4.3

	0	1	2	3	4	5	6	...	i	...	j	...
DIAG	1	0	0	1	0	0	0		0		1	...

Fig. 4.4

representation differs from each row of the table (Fig. 4.3). We show it by constructing a sequence of bits, called DIAG, that does not occur in any row of the table. The construction of the bit sequence DIAG and the corresponding set $M(\text{DIAG})$ is done using the diagonalization method.

First, see the binary value a_{00} in the intersection of the 0-th row and the 0-th column. If $a_{00} = 0$ (Fig. 4.3), i.e., if 0 does not belong to $M(P_0)$, then we set the 0-th position d_0 of DIAG to 1. If $a_{00} = 1$ (i.e., if 0 is in $M(P_0)$), then we set $d_0 = 0$ (i.e., we do not take 0 into $M(\text{DIAG})$). After this first step of the construction of DIAG we fixed only the value of the first position of DIAG, and due to this we are sure that DIAG differs from the 0-th row of the table (i.e., from $M(P_0)$), at least with respect to the membership of the 0-th element.

Analogously, we continue in the second construction step. We consider the second diagonal square, where the first row intersects the first column. We aim to choose the first position d_1 of DIAG in such a way that DIAG differs from the binary representation of $M(P_1)$ at least in the value of this position. Therefore, if $a_{11} = 1$ (i.e., if 1 is $M(P_1)$), we set d_1 to 0 (i.e., we do not take 1 into $M(\text{DIAG})$). If $a_{11} = 0$ (i.e., if 1 is not in $M(P_1)$), then we set $d_1 = 1$ (i.e., we take 1 into $M(\text{DIAG})$).

If $\bar{a}_{ij}$ represents the opposite value to a_{ij} for any bit in the intersection of the i-th row and the j-th column (the opposite value to 1 is the value $\bar{1} = 0$, and $\bar{0} = 1$ is the opposite value to 0), then, after two construction steps, we reach the situation as depicted in Fig. 4.5.

	0	1	2	3	4	$\cdots$	i	$i+1$	$\cdots$
DIAG	$\bar{a}_{00}$	$\bar{a}_{11}$	?	?	?	$\cdots$	?	?	$\cdots$

Fig. 4.5

The first two elements of DIAG are $\bar{a}_{00}$ and $\bar{a}_{11}$, and so DIAG differs from both $M(P_0)$ and $M(P_1)$. The remaining positions of DIAG are still not determined, and we aim to fix them in such a way that DIAG will differ from each row of the table in Fig. 4.3.

In general, we guarantee a difference between DIAG, and the i-th row of the table in Fig. 4.3 as follows. Remember that $\bar{a}_{ii}$ is the bit of the square in the intersection of the i-th row and the i-th column and that d_i denotes the i-th bit of DIAG. If $\bar{a}_{ii} = 1$ (i.e., if i belongs to $M(P_i)$), then we set $d_i = 0$ (i.e., we do not take i into $M(\text{DIAG})$). If $\bar{a}_{ii} = 0$ (i.e., if i is not in $M(P_i)$), then we set $d_i = 1$ (i.e., we take i into $M(\text{DIAG})$). Hence, $M(\text{DIAG})$ differs from $M(P_i)$.

	0	1	2	3	4	$\cdots$	i	$\cdots$
DIAG	$\bar{a}_{00}$	$\bar{a}_{11}$	$\bar{a}_{22}$	$\bar{a}_{33}$	$\bar{a}_{44}$	$\cdots$	$\bar{a}_{ii}$	$\cdots$

Fig. 4.6

By this approach DIAG is constructed in such a way that it does not occur in any row of the table. For the concrete, hypothetical table in Fig. 4.3, Fig. 4.4 shows the corresponding representation of DIAG. In general, one can outline the representation of DIAG as done in Fig. 4.6.

In this way we obtain the result that

> M(DIAG) *is not accepted by any program, and therefore the decision problem* ($\mathbb{N}$, M(DIAG)) *cannot be solved by any algorithm.*

One can specify M(DIAG) also in the following short way:

$$\begin{aligned} M(\text{DIAG}) &= \{n \in \mathbb{N} \mid n \text{ is not in } M(P_n)\} \\ &= \text{the set of all natural numbers } n, \\ &\quad\ \text{such that } n \text{ is not in } M(P_n). \end{aligned}$$

Exercise 4.7 Assume the intersection of the first 10 rows and the first 10 columns in the table of all programs contains values as written in Fig. 4.7. Estimate the first 10 positions of DIAG.

	0	1	2	3	4	5	6	7	8	9	⋯
$M(P_0)$	1	1	1	0	0	1	0	1	0	1	
$M(P_1)$	0	0	0	0	0	0	0	0	0	0	
$M(P_2)$	0	1	1	0	1	0	1	1	0	0	
$M(P_3)$	1	1	1	0	1	1	0	0	0	0	
$M(P_4)$	1	1	1	1	1	1	1	0	1	0	
$M(P_5)$	0	0	1	0	0	1	0	1	1	0	
$M(P_6)$	1	0	0	0	1	0	1	0	0	0	
$M(P_7)$	1	1	1	1	1	1	1	1	1	1	
$M(P_8)$	0	0	1	1	0	0	1	1	0	0	
$M(P_9)$	1	0	1	0	1	0	1	0	1	0	
$M(P_{10})$	0	0	1	0	0	0	1	1	0	1	
⋮											⋱

Fig. 4.7

Exercise 4.8 (challenge) Consider

$$M(\text{2-DIAG}) = \text{the set of all even numbers } 2i, \text{ such that } 2i \text{ is not in } M(P_i).$$

Is the decision problem ($\mathbb{N}$, M(2-DIAG)) algorithmically solvable? Carefully explain your argument! Draw diagrams that would, similarly to Fig. 4.3 and Fig. 4.4, show the construction of 2-DIAG.

Exercise 4.9 (challenge) Can you use the solution to Exercise 4.8 in order to define two other subsets of $\mathbb{N}$ that are not algorithmically solvable? How many algorithmically unsolvable problems can be derived by diagonalization?

Exercise 4.10 (challenge) Consider

> $M(\text{DIAG}_2)$ as the set of all even natural numbers $2i$ such that $2i$ is not in $L(P_{2i})$.

Can you say something about the algorithmic solvability of $(\mathbb{N}, M(\text{DIAG}_2))$?

Now, we know that the decision problem $(\mathbb{N}, M(\text{DIAG}))$ is not algorithmically solvable. But we are not satisfied with this result. The problem looks as if it could be described in a finite way by our construction, though it is represented by an infinite sequence of bits. But our construction does not provide any algorithm for generating DIAG because, as we will see later, though the table in Fig. 4.3 really exists, it cannot be generated by an algorithm. Moreover, the decision problem $(\mathbb{N}, M(\text{DIAG}))$ does not correspond to any natural task arising in practice.

4.4 Reduction Method or How a Successful Method for Solving Problems Can Be Used to Get Negative Results

We already know how to use the diagonalization method in order to describe algorithmically unsolvable problems. This provides a good starting position for us. In this section, we learn how to "efficiently" spread the proofs of algorithmic unsolvability to further problems. The main idea is to introduce the relation "easier or equally hard" or "not harder than" with respect to the algorithmic solvability.

Let U_1 and U_2 be two problems. We say that

> **U_1 is easier than or as hard as U_2**

or that

U_1 is not harder than U_2

with respect to algorithmic solvability, and we write

$$U_1 \leq_{Alg} U_2,$$

if the algorithmic solvability of U_2 implies (guarantees) the algorithmic solvability of U_1.

What does it mean exactly? If

$$U_1 \leq_{Alg} U_2$$

holds, then the following situations are possible:

- *U_1 and U_2 are both algorithmically solvable.*
- *U_1 is algorithmically solvable, and U_2 is not algorithmically solvable.*
- *Both U_1 and U_2 are not algorithmically solvable.*

The only situation that the validity of the relation $U_1 \leq_{Alg} U_2$ excludes is the following one[5]:

- *U_2 is algorithmically solvable and U_1 is not algorithmically solvable.*

Assume that the following sequence of relations

$$U_1 \leq_{Alg} U_2 \leq_{Alg} U_3 \leq_{Alg} \cdots \leq_{Alg} U_k$$

between the k problems $U_1, U_2, \ldots, U_k$ was proved. Moreover, assume that one can show by the diagonalization method that

U_1 is not algorithmically solvable.

What can be concluded from these facts? Since U_1 is the easiest problem among all problems of the sequence, all other problems $U_2, U_3, \ldots, U_k$ are at least as hard as U_1 with respect to algorithmic solvability, and so one can conclude that

[5] Remember the definition of the notion of implication in Chapter 1. The truthfulness of the implication "The solvability of U_2 implies the solvability of U_1" excludes exactly this one situation.

the problems $U_2, U_3, \ldots, U_k$ are not algorithmically solvable.

This is the way we want to walk around in order to prove the algorithmic unsolvability of further problems. Due to diagonalization we already have the initial problem U_1 for this approach. This problem is the decision problem $(\mathbb{N}, M(\text{DIAG}))$. The only question is, how to prove the validity of the relation $U_1 \leq_{Alg} U_2$ between two problems?

For this purpose, we apply the reduction method, which was developed in mathematics in order to solve new problems by clever application of known methods for solving other problems. We give two examples showing a simple application of the reduction method.

Example 4.1 Assume one has a method for solving quadratic equations in the so-called p, q-form

$$x^2 + px + q = 0,$$

i.e., quadratic equations with the coefficient 1 before the term x^2. The method for solving such quadratic equations is given by the so-called p-q-formula:

$$x_1 = -\frac{p}{2} + \sqrt{\left(\frac{p}{2}\right)^2 - q}$$

$$x_2 = -\frac{p}{2} - \sqrt{\left(\frac{p}{2}\right)^2 - q}.$$

If $\left(\frac{p}{2}\right)^2 - q < 0$ holds, then the quadratic equation in the p, q-form does not have any real solution.

Now, we are searching for a method for solving arbitrary quadratic equations

$$ax^2 + bx + c = 0\ .$$

Instead of deriving a new formula[6] for this purpose, we reduce the problem of solving general quadratic equations to the problem of solving quadratic equations in the p, q-form.

---

[6] We presented such a formula in Chapter 2 and wrote a program computing solutions using this formula there.

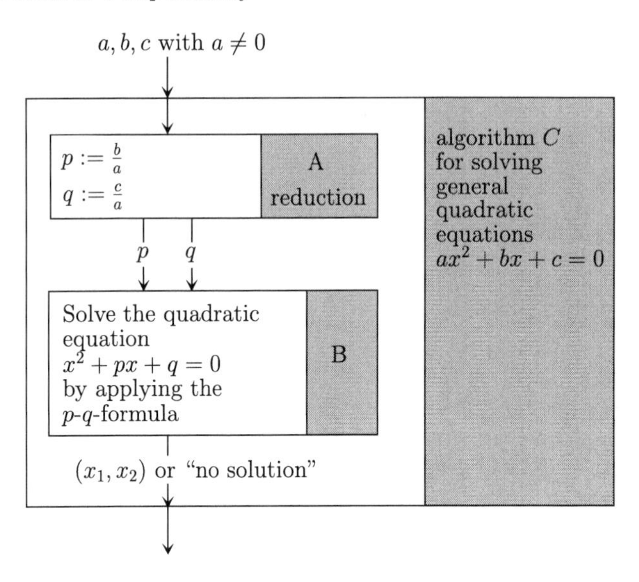

Fig. 4.8

We know that the solutions of an arbitrary equation do not change if one multiplies both sides of the equation by the same numbers apart from 0. Hence, we are allowed to multiply both sides of the general quadratic equation by $1/a$.

$$\begin{aligned} ax^2 + bx + c &= 0 \qquad\qquad \mid \cdot \frac{1}{a} \\ a \cdot \frac{1}{a} \cdot x^2 + b \cdot \frac{1}{a} x + c \cdot \frac{1}{a} &= 0 \cdot \frac{1}{a} \\ x^2 + \frac{b}{a} x + \frac{c}{a} &= 0. \end{aligned}$$

In this way we get a quadratic equation in the p, q-form and this equation can be solved using the method presented above. An algorithmic representation of this reduction is outlined in Fig. 4.8.

Part A is an algorithm that corresponds to the reduction. Here, one computes the coefficients p and q of the equivalent quadratic equation in the p, q-form. This is all one has to do in this algorithmic reduction. The coefficients p and q are the inputs for algorithm B for solving quadratic equations in the form $x^2 + px + q = 0$. B solves this equation for the given p and q. The output of B (either the solutions x_1 and x_2 or the answer "there is no solution") can used over as the output of the whole algorithm C for solving general quadratic equations. □

Exercise 4.11 Assume we have an algorithm B for solving linear equations in the form

$$ax + b = 0.$$

Design an algorithm for solving linear equations of the form

$$cx + d = nx + m$$

by reduction. The symbols c, d, n, and m stand for concrete numbers, and x is the unknown. Outline the reduction in a way similar to that used in Fig. 4.8.

The reduction form in Example 4.1 is called **1-1-reduction** (one to one reduction). It is the simplest possible reduction, in which the input of a problem U_1 (a general quadratic equation) is directly transformed to an input of a problem U_2 (a quadratic equation in the p, q-form), and the result of the computation of the algorithm on the input of U_2 is used one to one as the result for the given input instance of U_1. This means that

$$U_1 \leq_{Alg} U_2 \tag{4.1}$$

holds. In other words, solving U_1 in an algorithmic way is not harder than solving U_2, because each algorithm B solving U_2 can be "modified" by reduction (Fig. 4.8) to an algorithm C that solves U_1.

Moreover, in the case of quadratic equations, we observe that U_2 (solving quadratic equations in p, q-form) is a special case of U_1 (solving general quadratic equations). Hence, each algorithm for U_1 is automatically an algorithm for U_2, and so

$$U_2 \leq_{Alg} U_1. \tag{4.2}$$

Following Equations (4.1) and (4.2) we may claim that U_1 and U_2 are equally hard. This means that either both problems are algorithmically solvable or both are not algorithmically solvable. Clearly, we know in this special case of solving quadratic equations that the first possibility is true.

In general, the reductions need not be so simple as the one presented. To prove

$$U_1 \leq_{Alg} U_2$$

one can need to apply the algorithm B solving U_2 several times for different inputs and additionally to work on the outputs of B in order to compute the correct results for U_1. To illustrate such a more general reduction, we present the following example.

Example 4.2 We assume that everybody is familiar with the Pythagorean theorem, which says that in any right-angled triangle (Fig. 4.9) the equality

$$c^2 = a^2 + b^2$$

holds. In other words

> *The square of the length of the longest side of any right-angled triangle is equal to the sum of the squares of the lengths of the two shorter sides.*

In this way, we obtain an algorithm $B_\triangle$ that for given lengths of two sides of a right-angled triangle computes the length of the third side. For instance, for known a and b, one can compute c using the formula

$$c = \sqrt{a^2 + b^2}.$$

If a and c are known, one can compute b as

$$b = \sqrt{c^2 - a^2}.$$

Let us denote by $U_\triangle$ the problem of computing the missing side length of a right-angled triangle.

Assume now a new task U_{Area}. For a given equilateral triangle (Fig. 4.10) with side lengths m, one has to compute the area of the triangle. We see (Fig. 4.10) that the area of such a triangle is

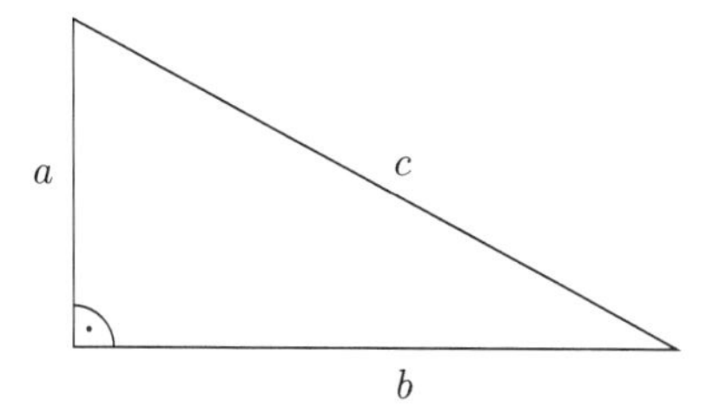

Fig. 4.9

$$\frac{m}{2} \cdot h$$

where h is the height of the triangle (Fig. 4.10).

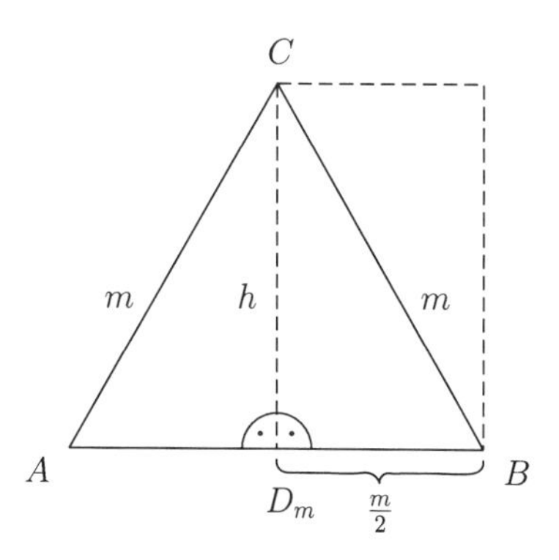

Fig. 4.10

We are able to show

$$U_{Area} \leq_{Alg} U_{\triangle}$$

by reducing solving U_{Area} to solving $U_{\triangle}$. How to do it is depicted in Fig. 4.11.

We designed an algorithm A_{Area} for solving U_{Area} under the assumption that one has an algorithm $B_{\triangle}$ for solving $U_{\triangle}$ (for computing the missing side length in a right-angled triangle). We see in Fig. 4.11 that one needs the height h of the triangle in order

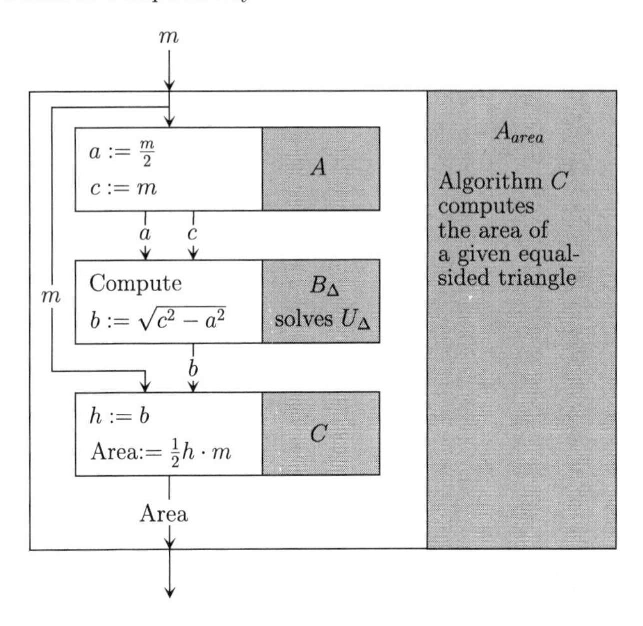

Fig. 4.11

to compute the area of the triangle. The height h is the length of the side CD of the right-angled triangle DBC. We also see that the length of the side DB is equal to m. Algorithm A in Fig. 4.11 uses these facts to compute the values a and c and use them as inputs for $B_\triangle$. The algorithm $B_\triangle$ computes the missing length b of $\triangle DBC$, which is the height h of $\triangle ABC$. Finally, the algorithm C computes the area of $\triangle ABC$ from the values of m and b. □

Exercise 4.12 Consider the problem U_{Pyr} of computing the height of a pyramid with quadratic base of size $m \times m$ and the lengths of the edges also equal to m. Solve this task by showing $U_{Pyr} \leq_{Alg} U_\triangle$. Show the reduction as we did for Fig. 4.11.

[Hint: Consider the possibility of applying the algorithm $B_\triangle$ twice for different inputs.]

Exercise 4.13 (challenge) Let U_{2lin} denote the problem of solving a system of two linear equations

$$a_{11}x + a_{12}y = b_1$$
$$a_{21}x + a_{22}y = b_2$$

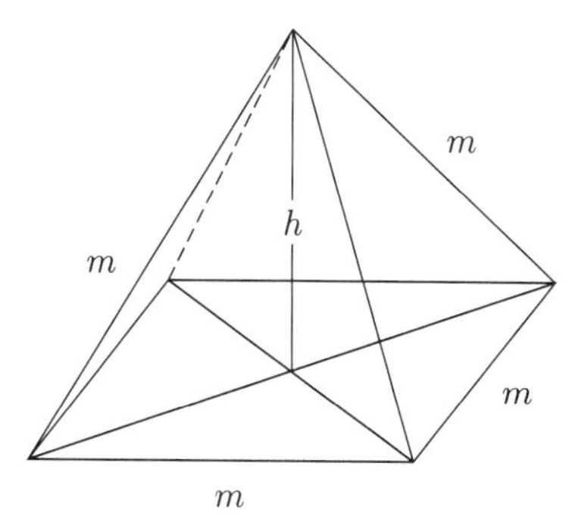

Fig. 4.12

with two unknown x and y. Let U_{3lin} denote the problem of solving a system of three linear equations

$$\begin{aligned} a_{11}x + a_{12}y + a_{13}z &= b_1 \\ a_{21}x + a_{22}y + a_{23}z &= b_2 \\ a_{31}x + a_{32}y + a_{33}z &= b_3 \end{aligned}$$

with three unknown x, y, and z. Show that $U_{3lin} \leq_{Alg} U_{2lin}$.

We saw how reductions can be used to develop new methods for solving problems by using known methods for solving other problems. In this way, one uses reduction to extend the positive results about algorithmic solvability.

But our aim is not to use reduction as a means for designing new algorithms (i.e., for broadcasting positive messages about algorithmic solvability). We aim to use reduction as an instrument for spreading negative messages about algorithmic unsolvability. How can one reverse a method for designing positive results to a method for proving negative results? We outlined this idea already at the beginning of this chapter. If one is able to prove

$$U_1 \leq_{Alg} U_2$$

by a reduction and one knows that U_1 is not algorithmically solvable, then one can conclude that U_2 is also not algorithmically solvable.

There is a simple difference between proving

$$U_1 \leq_{Alg} U_2$$

for extending algorithmic solvability to a new problem and for showing algorithmic unsolvability. For deriving a positive result, one already has an algorithm for U_2 and one tries to use it in order to develop an algorithm for U_1. To broaden negative results about unsolvability, we do not have any algorithm for U_2. We only *assume that there exists an algorithm solving* U_2. Under this assumption we build an algorithm that solves U_1. This means that we have to work with the hypothetical existence of an algorithm A_2 for U_2 and use it to design an algorithm for U_1.

Applying reduction to show the algorithmic solvability of a problem corresponds to a direct proof (direct argumentation) that was introduced in Chapter 1. Using reduction for proving the nonexistence of any algorithm for solving the problem considered corresponds to an indirect proof, whose schema was presented in Section 1.2. To get a transparent connection to something known, we give an example from geometry first, and then we switch to algorithmics.

Example 4.3 One knows that it is impossible to partition an arbitrary given angle into three equal-sided angles by means of a ruler and a pair of compasses. In other words, there is no method as a sequence of simple construction steps executable by means of a ruler and a pair of compasses that would guarantee a successful geometric partitioning of any angle into three equal-sided angles. The proof of this negative result is far from being obvious and so we prefer here to trust the mathematicians and believe it.

On the other hand, one may know that there are simple methods for geometric doubling or halving of each angle.

For instance, Fig. 4.13 outlines how to double the angle $\angle ab$ between two lines a and b. An RC-algorithm (a ruler–compass-algorithm) for doubling an angle can work as follows:

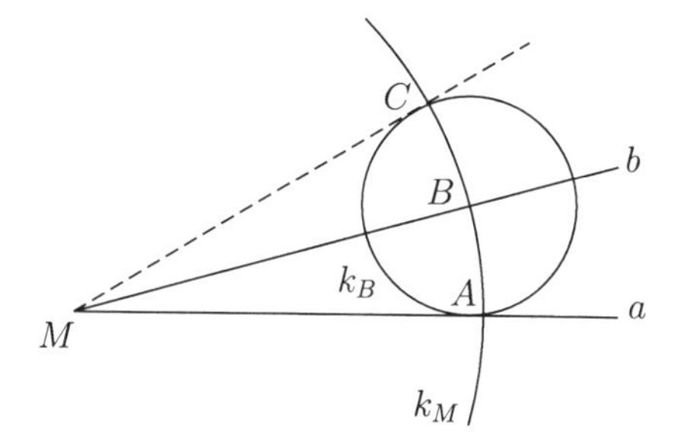

Fig. 4.13: Doubling an angle

1. Take an arbitrary positive distance r with the compass and draw a circle k_M with center M (the intersection of a and b) and radius r.
2. Denote by A the intersection of the circle k_m and the line a, and denote by B the intersection of k_M and the line b.
3. Take the distance $\overline{AB}$ between A and B using the compass and draw a circle k_B with center B and radius $\overline{AB}$.
4. Denote by C the intersection of the circles k_m and k_B that is different from A.
5. Connect the points M and C with a line.

We see that the angle $\angle AMC$ between the straight line a and the straight line connecting M and C is twice as large as the original angle $\angle ab = \angle AMB$.

Now, our aim is to show that there do not exist any RC-algorithms that can partition an arbitrary angle into 6 equal-sided parts. This does not look easier than to prove the nonexistence of any RC-algorithm for partitioning angles into three equal-sided angles. Our advantage is that we are not required to use this hard way of creating a nonexistence proof. We know already that one cannot partition angles into three equal-sided parts by means of a ruler and a compass. We use this fact to reach our aim efficiently.

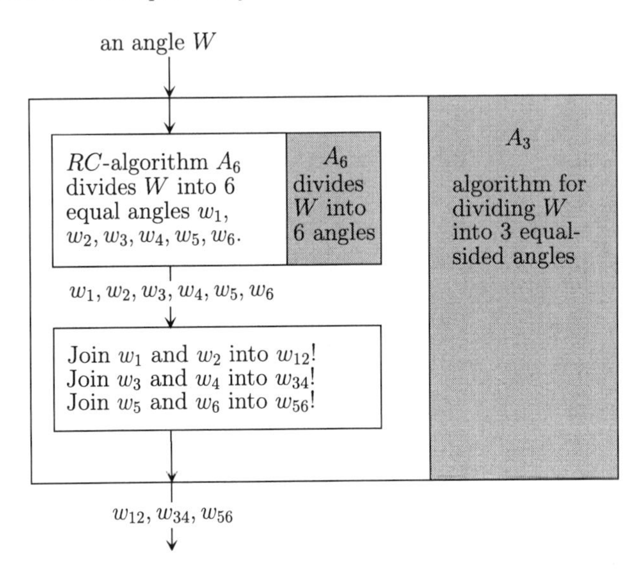

Fig. 4.14

How to proceed? We assume the opposite of what we want to prove and show under this assumption that one can partition angles into three equal-sided parts using a RC-algorithm. But this contradicts the already known fact about the impossibility of partitioning angles into three equal-sided parts. More precisely, we assume that there is a RC-algorithm A_6 for partitioning angles into 6 equal-sided angles and use A_6 to design a RC-algorithm A_3 that partitions each given angle into three equal-sided angles. Since A_3 does not exist (as we already know), A_6 cannot exist either.

We describe the reduction of the problem of partitioning angles into three equal-sided angles to the problem of partitioning angles into 6 equal-sided parts as follows (Fig. 4.14). We assume that one has a RC-algorithm A_6 for partitioning angles into 6 equal-sided parts. We design a RC-algorithm A_3 that has an angle W as its input. At the beginning, A_3 applies A_6 in order to partition W into 6 equal-sided angles $w_1, w_2, w_3, w_4, w_5, w_6$ (Fig. 4.15). Then, A_3 joins the angles w_1 and w_2 into the angle w_{12}. Similarly, A_3 joins w_3 and w_4 into an angle w_{34}, and w_5 and w_6

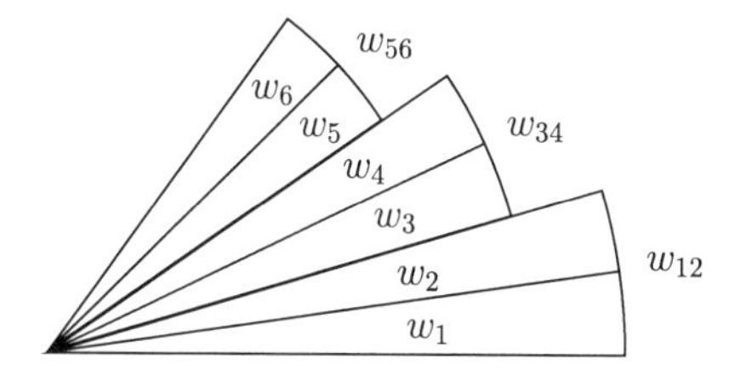

Fig. 4.15

are joined into w_{56} (Fig. 4.15). We see that partitioning W into three angles w_{12}, w_{34}, and w_{56} corresponds to the partitioning of W.

Using the language of the indirect argumentation (indirect proof), the reduction in Fig. 4.14 corresponds to the following implication:

> *If there is a RC-algorithm for partitioning angles into 6 equal-sided parts, then there is a RC-algorithm for partitioning angles into 3 equal-sided angles.*

Following the definition of the notion "implication", the truthfulness of this implication proved above (Fig. 4.14) excludes the second situation from the 4 possible situations listed in Fig. 4.16.

situation	6 parts	3 parts
1	possible	possible
2	possible	impossible
3	impossible	possible
4	impossible	impossible

Fig. 4.16

Taking into account that partitioning angles into three equal-sided parts is impossible, situations 1 and 3 are excluded, too. Hence, the only remaining possible situation is situation 4. Situation 4 contains the impossibility of partitioning angles into 6 equal-sided parts, and this completes our indirect argumentation for the nonexistence of a RC-algorithm partitioning angles into 6 equal-sided angles. □

Exercise 4.14 The problem of partitioning of an angle into three parts also has the following simplified representation. The task is to construct, for any given angle W, an angle V by means of a ruler and a compass such that the size of V is one third of the size of W. One can prove that this simplification does not change anything about the RC-unsolvability of this problem. Use this fact to create in a similar way as in Fig. 4.14 reductions (proofs) showing the nonexistence of RC-algorithms for constructing angles of the size of

(i) one sixth
(ii) one ninth

of any given angle.

Now, we return from the world of RC-algorithms into the world of general algorithms. Our problem DIAG plays here a similar role as the problem of partitioning angles into three equal-sided parts does for RC-algorithms. Starting from algorithmic unsolvability of a problem, we want to conclude algorithmic unsolvability of another problem.

The reduction schema for $U_1 \leq_{Alg} U_2$ is outlined in Fig. 4.17.

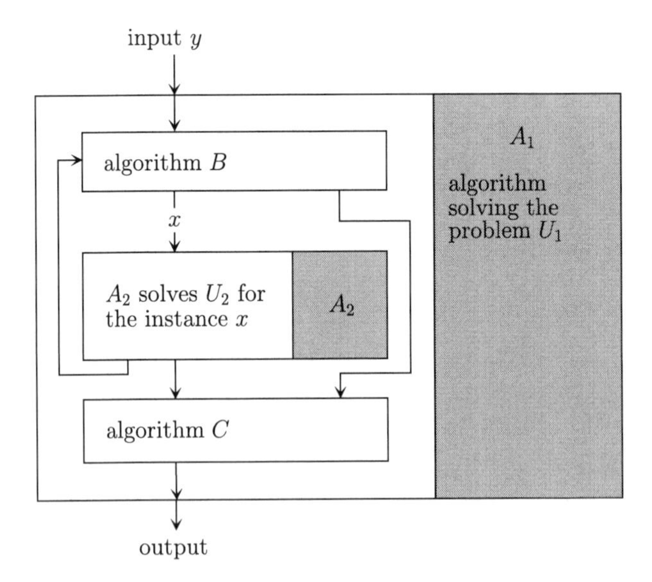

Fig. 4.17

The algorithm A_1 solving the problem U_1 is created as follows. First, the input instance y of U_1 is processed by an algorithm B that transforms y into a problem instance x of the problem U_2. Following our assumption about the existence of an algorithm A_2 solving U_2, the algorithm A_2 computes the correct solution for the input x. As one may see in Fig. 4.17, A_2 can be used repeatedly several times. Finally, an algorithm C processes all outputs of A_2 and computes the final result for the problem instance y of U_1. We call attention to the important fact that A_2, B, and C are algorithms and therefore they provide their outputs in a final time. The number of requests on A_2 must be finite and so the loop containing B and A_2 can run only finitely many times. Therefore, we can conclude that A_1 is an algorithm for U_1 because it provides the correct output in a finite time for any input instance of U_1.

Next, we introduce two new decision problems that are of interest for program developers.

UNIV (the **universal problem**)
Input: a program P and an input $i \in \mathbb{N}$ for P
Output: YES, if P accepts the input i, i.e., if i is in $M(P)$.
NO, if P does not accept i
(i.e., if $i \notin M(P)$), which means that P either halts and rejects i or P works infinitely long on the input i.

HALT (the **halting problem**)
Input: a program P and a natural number i
Output: YES, if P halts on the input i
(i.e., if P finishes its work on i in a finite time).
NO, if P does not halt on i
(i.e., if P has an infinite computation on i repeating a loop infinitely many times).

The halting problem is one of the fundamental tasks in testing software products. We already know that only those programs can be considered to be algorithms that never get into an infinite computation. Hence, an important part of checking the correct functionality of programs is to verify whether they always (for every input) guarantee an output in a finite time. The halting problem HALT is a simple version of such testing. We are only asking whether a given program P halts on a concrete input i. (The real question is whether a given program halts on every possible input.) Later, we will see that even this simplified testing problem is algorithmically unsolvable.

The universal problem UNIV is directly related to verifying the correct functionality of a program solving a decision problem. We test whether P provides the correct result YES or NO on an input i. Now, somebody can propose the following simple way to solve UNIV. Simulate the work of P on i and look at whether P outputs YES or NO. Certainly, one can do it if one has a guarantee that P halts on i (i.e., that P is an algorithm). But we do not have this guarantee. If P executes an infinite computation on i, we would simulate the work of P on i infinitely long and would never get the answer to our question of whether P accepts i or not. An algorithm for the universal problem is not allowed to work infinitely long on any input P and i, and so it is not allowed to execute an infinite simulation.

Following these considerations, we get the impression that the halting problem and the universal problem are strongly interlocked. Really, we show that these problems are equally hard.

First we show that

$$\text{UNIV} \leq_{Alg} \text{HALT}$$

i.e., that

UNIV is not harder than HALT with respect to algorithmical solvability.

What do we have to show? We have to show that the existence of an algorithm for HALT assures the existence of an algorithm that solves UNIV. Following our schema of indirect proofs, we assume that one has an algorithm A_{HALT} solving the halting problem. Using this assumption, we build an algorithm B that solves UNIV (Fig. 4.18).

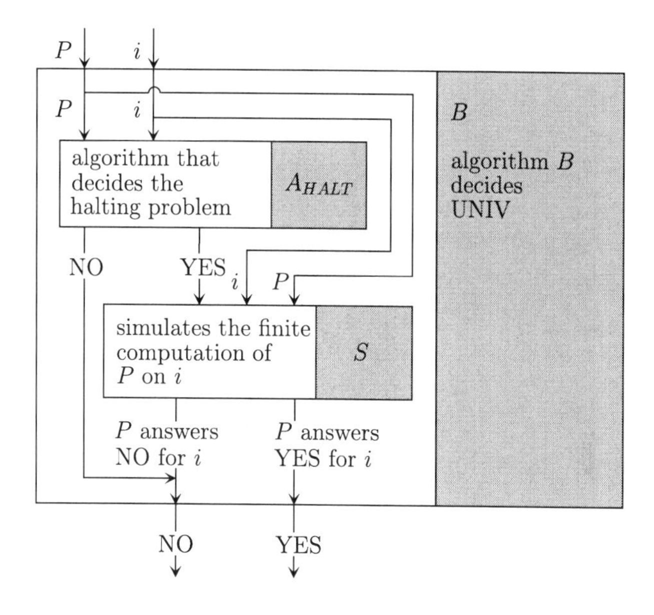

Fig. 4.18

The algorithm B works on any input (P, i) as follows:

1. B transfers its input (P, i) without any change to the algorithm A_{HALT}.
2. The algorithm A_{HALT} decides (in finite time) whether P halts on i or not. A_{HALT} answers YES if P halts on i. If P does not hold on i, A_{HALT} answers NO.
3. If A_{HALT} outputs NO, B is sure about the fact that P does not accept i (because P works on i infinitely long) and provides the final output NO saying that "i is not in $M(P)$".
4. If A_{HALT} outputs YES, then B simulates using a subprogram S (Fig. 4.18) the finite computation of P on i. Executing this finite simulation B sees whether P accepts i or not and outputs the corresponding claim.

Following the construction of B, we immediately see that B takes the right decision with respect to the membership of i in $M(P)$. We still have to verify whether B always finishes its work in a finite time. Under the assumption that A_{HALT} is an algorithm for

HALT, we know that A_{HALT} provides outputs in a finite time, and so B cannot run for an infinitely long time in its part A_{HALT}. The simulation program S starts to work only if one has the guarantee that the computation of P on i is finite. Therefore, the simulation runs always in a finite time, and hence B cannot get into an infinite repetition of a loop in the part S. Summarizing, B always halts, and so B is an algorithm that solves the universal problem.

We showed above that UNIV is easier than or equally hard as HALT. Our aim is to show that these problems are equally hard. Hence, we have to show that the opposite relation

$$\text{HALT} \leq_{Alg} \text{UNIV}$$

holds, too. This means, we have to show that the algorithmic solvability of UNIV implies the algorithmic solvability of HALT. Let A_{UNIV} be an algorithm that decides UNIV. Under this assumption we design an algorithm D that solves HALT. For any input (P, i), the algorithm D works as follows (Fig. 4.19).

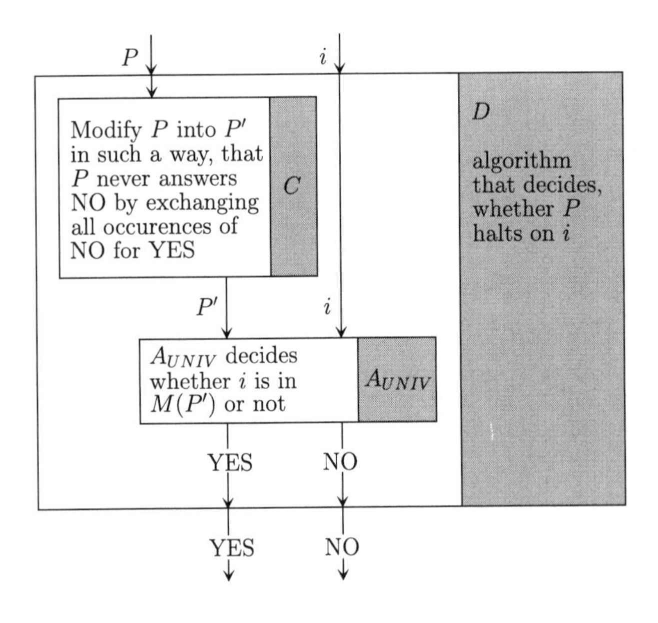

Fig. 4.19

1. D gives P to a subprogram C that transforms P into P' in the following way. C finds all rows of P containing the instruction *output ("NO")* and exchanges "NO" for "YES". Hence, the constructed program never outputs "NO" and the following is true:

   Each finite computation of P finishes with the output YES and P' accepts exactly those natural numbers i on which P halts.

2. D gives P' and i as inputs to A_{UNIV} (Fig. 4.19). A_{UNIV} decides whether i is in $M(P')$ or not.

3. D takes over the answer YES or NO of A_{UNIV} as the final answer for its input (P, i).

Exercise 4.15 Provide a detailed explanation for why D is an algorithm that solves the halting problem.

Exercise 4.16 (challenge) The reduction for $A_{\text{UNIV}} \leq_{Alg} A_{\text{HALT}}$ in Fig. 4.18 and the reduction $A_{\text{HALT}} \leq_{Alg} A_{\text{UNIV}}$ (Fig. 4.19) look different. Usually, one prefers the kind of reduction presented in Fig. 4.19 because it corresponds to the typical reduction in mathematics. Here, one transforms the input instance (P, i) of HALT to an input instance (P', i) of UNIV in such a way that the solution for the instance (P, i) of HALT is exactly the same as the solution for (P', i) of UNIV. Due to this, one can take the answer of A_{UNIV} for (P', i) as the final output for the input (P, i) of HALT. The schema of this reduction is the simple schema presented in Fig. 4.8 and Fig. 4.19. Find such a simple reduction for the proof of $A_{\text{UNIV}} \leq_{Alg} A_{\text{HALT}}$. This means you have to algorithmically transform the instance (P, i) of UNIV into such an instance (P', i) that the output of A_{HALT} for (P', i) (i.e., the reduction for (P', i) of the halting problem) corresponds to the solution for the instance (P, i) of UNIV.

Above, we showed that the universal problem and the halting problem are equally hard with respect to algorithmic solvability. This means that either both problems are algorithmically solvable or both are algorithmically unsolvable. As we already mentioned, we aim to prove their unsolvability. To do that, it is sufficient to show that one of them is not easier than $(\mathbb{N}, M(\text{DIAG}))$. Here, we prove

$$(\mathbb{N}, M(\text{DIAG})) \leq_{Alg} \text{UNIV}.$$

We assume the existence of an algorithm A_{UNIV} for UNIV and use A_{UNIV} to create an algorithm A_{DIAG} that decides $(\mathbb{N}, M(\text{DIAG}))$. For any natural number i, the algorithm A_{DIAG} has to compute the answer YES if the i-th program P_i does not accept the number i and the answer NO if P_i accepts i.

For any input i, our hypothetical algorithm A_{DIAG} works on i as follows (Fig. 4.20):

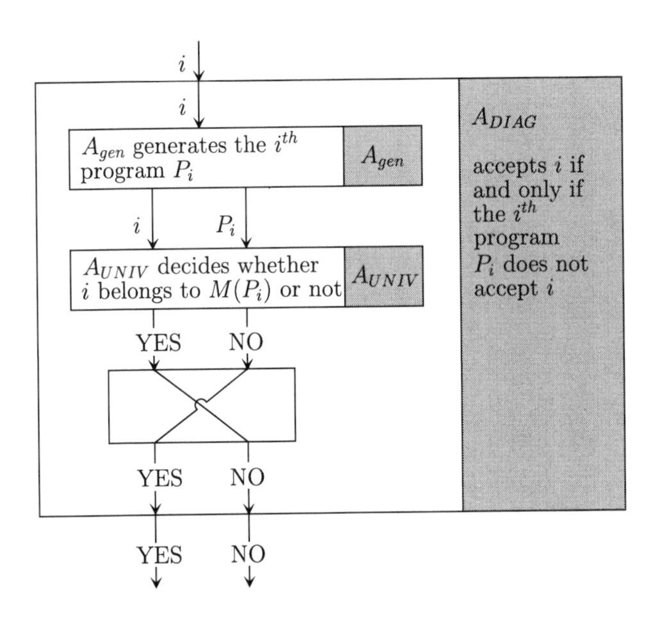

Fig. 4.20

1. A_{DIAG} gives i to a subprogram A_{gen} that generates the i-th program P_i.
2. A_{DIAG} gives i and P_i as inputs to the hypothetical algorithm A_{UNIV}. A_{UNIV} decides whether P_i accepts i (providing the answer YES) or not (providing the answer NO).
3. A_{DIAG} exchanges the answers of A_{UNIV}. If A_{UNIV} outputted YES (i.e., i is in $M(P_i)$), then i does not belong to $M(\text{DIAG})$ and A_{UNIV} computes NO. If A_{UNIV} computed NO (i.e., i is not in $M(P_i)$), then i is in $M(\text{DIAG})$ and A_{UNIV} correctly answers YES (Fig. 4.20).

Following the description of the work of A_{DIAG} on i, we see that A_{DIAG} provides the correct answers under the assumption that A_{UNIV} and A_{gen} work correctly. The fact that A_{UNIV} is an algorithm solving UNIV is our assumption. The only open question is whether one can build an algorithm A_{gen} that, for each given natural number i, constructs the i-th program P_i in a finite time. A_{gen} can work as follows (Fig. 4.21). It generates texts over the keyboard alphabet one after the other in the order described at the beginning of this chapter. For each text generated, A_{gen} uses a compiler in order to check whether the text is a program representation or not. Moreover, A_{gen} counts the number of positive compiler answers. After A_{gen} gets i positive answers, it knows that the last text generated is the i-th program P_i. The structure (flowchart) of the algorithm A_{gen} is outlined in Fig. 4.21.

Exercise 4.17 Show that $(\mathbb{N}, M(\text{DIAG})) \leq_{Alg}$ HALT is true by a reduction from $(\mathbb{N}, M(\text{DIAG}))$ to HALT.

Exercise 4.18 Let $M(\overline{\text{DIAG}})$ be the set of all natural numbers i, such that P_i accepts i. Hence, $M(\overline{\text{DIAG}})$ contains exactly those natural numbers that do not belong to $M(\text{DIAG})$. Show by a reduction that

$$(\mathbb{N}, M(\text{DIAG})) \leq_{Alg} (\mathbb{N}, M(\overline{\text{DIAG}})) \text{ and } (\mathbb{N}, M(\overline{\text{DIAG}})) \leq_{Alg} (\mathbb{N}, M(\text{DIAG})).$$

We showed that the decision problem $(\mathbb{N}, M(\text{DIAG}))$, the universal problem UNIV, and the halting problem HALT are not algorithmically solvable. The problems UNIV and HALT are very important for testing of programs and so are of large interest in software engineering. Unfortunately, the reality is that all important tasks related to program testing are algorithmically unsolvable. How bad this is can be imagined by considering the unsolvability of the following simple task.

Let f_0 be a function from natural numbers to natural numbers that is equal to 0 for every input $i \in \mathbb{N}$. Such functions are called constant functions because the result 0 is completely independent of the input (of the argument). The following program

```
0 Output ← ''0''
1 End,
```

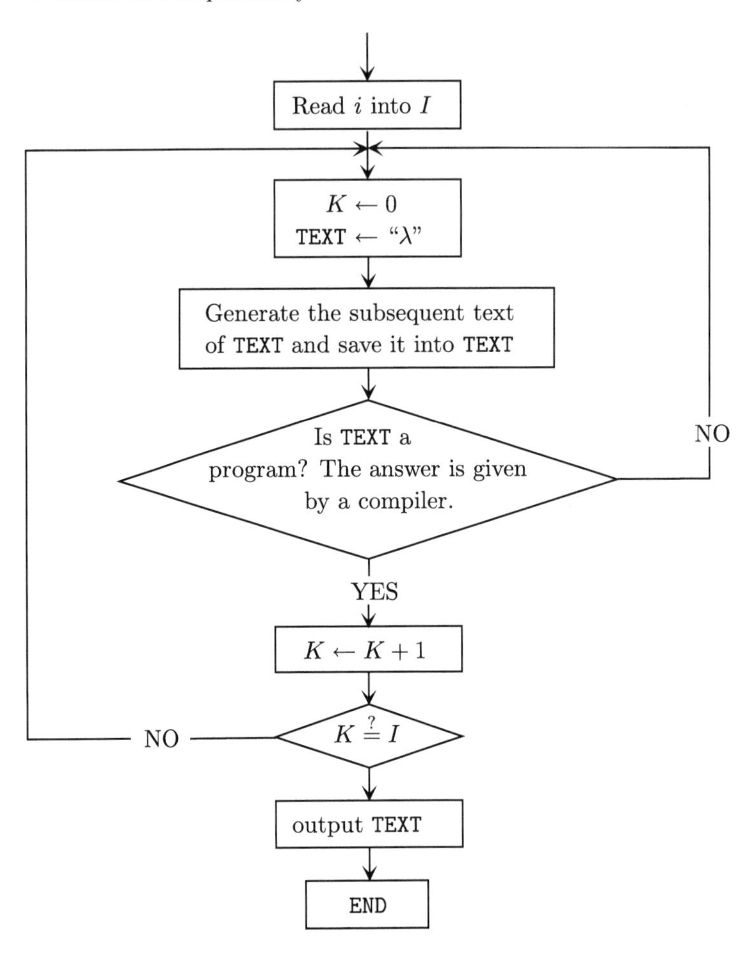

Fig. 4.21

computes $f_0(i)$ for each i, and we see that it does not need to read the input value i because i does not influence the output. Despite the simplicity of this task, there is no algorithm that is able to decide for a given program P whether P computes f_0 or not (i.e., whether P is a correct program for computing f_0). To understand it properly, note that input instances of this problem may also be very long complex programs doing additionally a lot of pointless work. The question is only whether at the end the correct result "0" is computed or not.

Exercise 4.19 (challenge) Let $\mathcal{M}_0$ be the set of all programs with $M(P) = \emptyset$. In other words, $\mathcal{M}_0$ contains all programs that, for each input, either output NO or execute an infinite computation. Prove that it is not algorithmically decidable whether a given program P belongs to $\mathcal{M}_0$ (whether P does not accept any input) or not.

In this chapter, we learned something very important. *All syntactic questions and problems such as "Is a given text a program representation" are algorithmically decidable.* We are even able to construct the i-th program P_i for any given i. *All the semantic questions and problems about the meaning and the correctness of programs are not algorithmically solvable.*

4.5 Summary of the Most Important Discoveries

We destroyed the dream from the early 20th century about automating everything. We discovered *the existence of tasks that cannot be automatically solved by means of computers controlled by algorithms.* This claim is true independently of current and future computer technologies.

Among the algorithmically unsolvable problems, one can find many tasks of interest such as

- Is a program correct? Does it fit the aim for which it was developed?
- Does a program avoid infinite computations (endless repetitions of a loop)?

In computer science, there are several huge research communities focusing on testing programs.[7] Unfortunately, even such *simple questions as whether a given program computes a constant function are not algorithmically solvable.* The scientists in this area are satisfied if they are able to develop algorithms for testing some kinds of partial correctness of programs. For instance, they try to

[7] This witnesses the importance of program testing in practice.

develop automatic testing of programs in some very special representation or at least to find some typical errors without any guarantee that all errors have been found.

For computing tasks related to program testing, one distinguishes between syntactic and semantic problems. **Syntactic** tasks are usually related to the correct representation of a program in the formalism of a given programming language, and these problem settings are algorithmically solvable. **Semantic** questions are related to the meaning of programs. For instance:

- What does a given program compute? Which problem does the program solve?
- Does the program developed solve the given problem?
- Does a given program halt on a given input or does a program always (for all inputs) halt?
- Does a given program accept a given input?

All nontrivial semantic tasks about programs are not algorithmically solvable.

To discover the above presented facts, we learned two proof techniques that can also be viewed as research methods. The first technique was the diagonalization method that we already applied in the study of infinities in Chapter 3. Using it, we were able to show that the number of problems is a larger infinity than the number of programs. Consequently, there exist algorithmically unsolvable problems. The first algorithmically unsolvable problem we discovered was the decision problem $(\mathbb{N}, M(\text{DIAG}))$. To extend its algorithmic unsolvability to other problems of practical interest, we used the reduction method. This method was used for a long time in mathematics in order to transform algorithmic solvability from problem to problem (i.e., for broadcasting positive messages about automatic solvability). The idea of the reduction method is to say that

P_1 is not harder than P_2, $P_1 \leq_{Alg} P_2$,

if, assuming the existence of an algorithm for P_2, one can build an algorithm solving P_1. If this is true, we say that P_1 *is reducible to* P_2.

The positive interpretation of the fact $P_1 \leq_{Alg} P_2$ is that algorithmic solvability of P_2 implies the algorithmic solvability of P_1. The negative meaning of $P_1 \leq_{Alg} P_2$ we used is that the algorithmic unsolvability of P_2 follows from the algorithmic unsolvability of P_1. We applied the reduction method in order to derive the algorithmic unsolvability of the halting problem and of the universal problem from the algorithmic unsolvability of the diagonal problem.

Solutions to Some Exercises

Exercise 4.3 The number of real numbers with a finite representation is exactly $|\mathbb{N}|$. We know that $2 \cdot |\mathbb{N}| = |\mathbb{N}|$ and even that $|\mathbb{N}| \cdot |\mathbb{N}| = |\mathbb{N}|$. Since $|\mathbb{R}| > |\mathbb{N}|$, we obtain

$$|\mathbb{R}| > |\mathbb{N}| \cdot |\mathbb{N}| .$$

Therefore, we see that the number of real numbers with a finite representation is an infinitely small fraction of the set of all real numbers.

Exercise 4.6 One can see the first 17 positions of the binary representation of QUAD in the following table:

	0	1	2	3	4	5	6	7	8	9	10	11	12	13	14	15	16	17
QUAD	1	1	0	0	1	0	0	0	0	1	0	0	0	0	0	0	1	0

You can easily extend this table for arbitrarily many positions.

Exercise 4.7 The first 10 positions of DIAG for the hypothetical table in Fig. 4.7 are

$$\text{DIAG} = 0101000011.$$

Exercise 4.9 We aim to show that the set

M(2-DIAG) = the set of all even positive integers $2i$, such that $2i$ is not in $M(P_i)$

is not decidable. The idea is again based on diagonalization as presented in Fig. 4.3. We create 2-DIAG in such a way that M(2-DIAG) differs from each row of the table. The only difference to constructing DIAG is that 2-DIAG differs from the i-th row of the table in position $2i$ (instead of i in the case of DIAG). The following table in Fig. 4.22 provides a transparent explanation of the idea presented above.

	0	1	2	3	4	5	6	7	8	9	10	11	12	...
M(P_0)	[0]	0	1	1	0	1	1	0	1	1	1	1	0	
M(P_1)	1	0	[1]	1	0	0	0	0	1	0	1	1	0	
M(P_2)	1	1	1	1	[1]	1	1	1	0	0	0	1	0	
M(P_3)	0	1	0	1	0	0	[0]	0	1	1	1	0	0	
M(P_4)	1	0	1	0	1	0	1	0	[1]	0	1	0	1	
M(P_5)	0	1	0	1	1	0	0	1	0	1	[0]	1	1	
M(P_6)	0	0	0	0	0	0	0	0	0	0	0	0	[0]	...
⋮													⋮	⋱

Fig. 4.22

The positions in boxes correspond to the intersections of the i-th row with the $2i$-th column. The set 2-DIAG differs from the rows of the table at least in the values written on these positions. Hence, the first 13 positions of 2-DIAG with respect to the hypothetical table in Fig. 4.22 are

$$\text{2-DIAG} = \underline{1}0\underline{0}0\underline{0}0\underline{1}0\underline{0}0\underline{1}0\underline{1}\ldots$$

We see that we took 0 for each odd position of 2-DIAG, and we note that the values of these positions do not have any influence on the undecidability of 2-DIAG. The underlined binary values starting with the 0-th position are related to the values in the boxes in the table in Fig. 2.2. In this way, the underlined bit 1 on the 0-th position of 2-DIAG guaranteed that 2-DIAG is different from the 0-th row of the table. The underlined bit 0 on the second position guarantees that 2-DIAG does not lie in the second row, etc. Bit 1 on the 12-th position of 2-DIAG ensures that 2-DIAG differs from the 6-th row.

Exercise 4.17 We aim to show that having an algorithm A_{HALT} solving HALT, one can algorithmically recognize the set $M(\text{DIAG})$. We start in a similar way as in Fig. 4.20, where the reduction $(|\mathbb{N}|, M(\text{DIAG})) \leq_{Alg}$ UNIV was outlined. Our task is to decide for a given i whether i belongs to $M(\text{DIAG})$ (i.e., whether P_i does not accept i) or not. We again use A_{gen} in order to generate the i-th program P_i and ask the algorithm A_{HALT} whether P_i halts on i or not (Fig. 4.23).

If P_i does not halt on i, then i does not belong to $M(P_i)$ and one knows with certainty that $i \notin M(P_i)$ (P_i does not accept i), and so that the correct answer is YES ("$i \in M(\text{DIAG})$"). If P_i halts on i, then we continue similarly as in the reduction presented in Fig. 4.18. We simulate the work of P_i on i in a finite time and convert the result of the simulation. If P_i does not accept i, then we accept i. If P_i accepts i, then we do not accept the number i (Fig. 4.23).

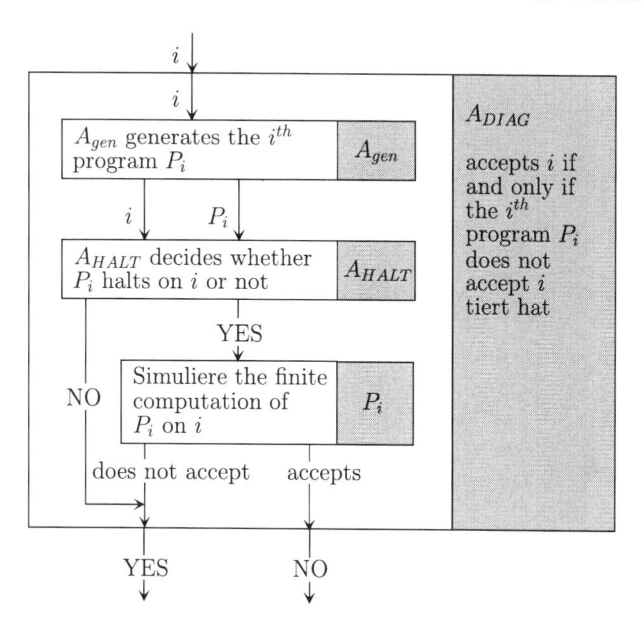

i
i
A_{gen} generates the ith program P_i
A_{gen}
i
P_i
A_{HALT} decides whether P_i halts on i or not
A_{HALT}
YES
NO
Simuliere the finite computation of P_i on i
P_i
does not accept
accepts
YES
NO
A_{DIAG}
accepts i if and only if the ith program P_i does not accept i
tiert hat

Fig. 4.23

There is no greater loss than time which has been wasted
Michelangelo Buonarroti

Chapter 5

Complexity Theory or What to Do When the Energy of the Universe Doesn't Suffice for Performing a Computation?

5.1 Introduction to Complexity Theory

In Chapter 4, we recognized that there exist interesting computing tasks that cannot be solved by means of algorithms. Moreover, we learned how to show that some problems are not algorithmically solvable. Until the early 1960s, the classification of computing problems into algorithmically solvable and algorithmically unsolvable dominated the research on fundamentals. The situation changed when more and more computers were used for civil purposes. Computers started to be frequently used for scheduling and optimization of working processes in industry and also to simu-

J. Hromkovič, *Algorithmic Adventures*, DOI 10.1007/978-3-540-85986-4_5,

late expensive research experiments. Trying to write programs for such purposes, the algorithm designers and programers recognized that many of the tasks to be solved by programmed computers are real challenges. Algorithms were designed, programs were implemented, computers ran and ran them, and all were covered in sweat in the computer room because the cooling of the processors was a serious problem at that time. The only problem was that the expected results of the computer work were not in view. At that time, the computers had to be frequently maintained, and so the computing time was restricted to the interval between two services. This interval did not suffice to successfully finish the computations. Computer scientists were not able to solve several problems in spite of the fact that these problems were algorithmically solvable. They were able to design algorithms for given problems and implement them, but not to use them in order to solve the problems. One saw the necessity of predicting the working time of algorithms designed. Therefore, the fundamental processes of building paradigmatic notions were activated once again. The notions of computational complexity of algorithms and, to some extent, of algorithmic problems were introduced. Soon, computer scientists started to imagine what the meaning of efficient computations really is. Many algorithms were not applicable not only because they were not able to finish their computations in a few days, but because the age of the Universe was too short for performing them. Hence, such algorithms were not useful. One could say, "OK, let us look for more efficient algorithms for the problems considered." The trouble was that there were several hundreds of problems (computing tasks) for which, in spite of a big effort, nobody was able to find an efficient algorithm. Again, principal question were posed:

Are we unable to find an efficient way for solving some problems because our knowledge in algorithmics is too poor?

Or do there exist algorithmically solvable problems that do not admit any efficient algorithm to solve (i.e., do there exist problems that are not practically solvable in spite of the fact they are algorithmically solvable)?

These questions led to the development of complexity theory, the aim of which is to measure the amount of computer resources needed for solving concrete problems, and so to classify algorithmic tasks with respect to their computational hardness. The main goal of complexity theory is to partition algorithmic problems into the class of practically solvable (tractable) problems and the class of intractable (practically unsolvable) problems. Complexity theory showed that there exist such computing problems that the whole energy of the Universe does not suffice for computing their solutions. Thus, we learned the following:

Algorithmically solvable does not mean tractable (practically solvable).

To recognize which problems are tractable and to develop efficient algorithms for tractable problems is, up to now, one of the hardest research topics in computer science.

The proofs and arguments in complexity theory are typically so involved and deep that all we have mastered up to now looks like a children's game in comparison with the considerations related to the existence of efficient algorithms. Therefore, we do not go into technical detail in this chapter. Fortunately, we do not need to present any involved technicalities in order to imagine the miracles we aim to introduce. The only thing we need is to understand some of the concepts and discoveries of complexity theory, this is the aim of this chapter.

5.2 How to Measure Computational Complexity?

The notion of complexity in the sense of the amount of computer work is central to computer science and the most important notion of computer science after the notions of algorithm and program. If one works in a precise mathematical way, one has to first agree on a mathematical model of algorithms and then to measure complexity as the number of computer model instructions performed. Fortunately, this precise measurement is necessary only for deriving the quantitative laws of information processing; their presentation is omitted here because of their high degree of technicality. For the usual work with algorithms, devoted to their design and implementation, the following simple approach for complexity measurement provides sufficiently trustable statements.

How can one measure the complexity of an algorithm in a simple way? Let A be an algorithm that solves a problem U. First we have to say what the complexity of A on a problem instance I is. The simplest kind of measurement defines **the complexity of *A* on *I*** as

> *the number of computer instructions performed in the computation of A on I.*

Because one assumes that the computer instructions are executed one after each other, one speaks about the **time complexity of *A* on *I***. Since the time complexity is the most important measure for judging the efficiency of algorithms, we often use the term complexity instead of time complexity in what follows. The second most important complexity for us is the **space complexity of *A* on *I***, which is

> *the number of variables and so the number of registers (memory units) used in the computation of A on I.*

For instance, consider the task of computing the value of a quadratic polynomial

$$a \cdot x^2 + b \cdot x + c$$

for given integers

$$a = 3, b = 4, c = 5, \text{ and } x = 7 .$$

A naive algorithm can work as follows:

$L \leftarrow b \cdot x$

{Multiply b by x and save the result in the variable (register named by) L}

$X \leftarrow x \cdot x$

$Y \leftarrow a \cdot X$

{After that the value of ax^2 is saved in Y}

$D \leftarrow L + c$

{After that the value $b \cdot x + c$ is saved in D}

$R \leftarrow Y + D$

{After that the result $ax^2 + bx + c$ is saved in R}

We immediately see that, for given numbers a, b, c, and x, the following 5 instructions (arithmetic operations) are executed:

$$\begin{array}{ccccc} b \cdot x \rightarrow L \ , & x \cdot x \rightarrow X \ , & a \cdot X \rightarrow Y \ , & L + c \rightarrow D \ , & Y + D \rightarrow R \\ 4 \cdot 7 = 28 \ , & 7 \cdot 7 = 49 \ , & 3 \cdot 49 = 147 \ , & 28 + 5 = 33 \ , & 147 + 33 = 180 \end{array}$$

Hence, the time complexity of A on $I = (a = 3, b = 4, c = 5, x = 7)$ is exactly 5. If one uses different registers for saving the values of a, b, c, x, L, X, Y, D, and R, then the space complexity of A on I is exactly 9. We observe that the values of a, b, c, and x do not have any influence on the complexity of A. Because of that we say that the time complexity of A is equal to 5 for any problem instance (for any quadratic polynomial).

Exercise 5.1 Transform the algorithm A into a program in the programming language of Chapter 2 that allows simple machine instructions only and forces us also to think about reading the values of a, b, c, and x.

(a) What is the time complexity of A in your implementation in this simple programming language?
(b) Can you rewrite the program in such a way that it uses fewer than 9 registers?

The procedure of the algorithm A can be made transparent by using the following representation of the polynomial:

$$a \cdot x \cdot x + b \cdot x + c \ .$$

Here, one immediately sees the three multiplications and two additions that have to be performed. Next, we try to improve the algorithm. Applying the well-known distributive law, we obtain

$$a \cdot x \cdot x + b \cdot x = (a \cdot x + b) \cdot x \ ,$$

and so the following representation of any quadratic polynomial:

$$ax^2 + bx + c = (ax + b) \cdot x + c \ .$$

This new representation contains only two multiplications and two additions, and so the complexity of the corresponding algorithm is 4.

Exercise 5.2 Consider a polynomial of degree four:

$$f(x) = a_4 \cdot x^4 + a_3 \cdot x^3 + a_2 \cdot x^2 + a_1 \cdot x + a_0 \ .$$

Explain how one can compute the value of the polynomial for given a_4, a_3, a_2, a_1, a_0, and x by 4 multiplications and 4 additions only.

Exercise 5.3 (challenge) Design an algorithm that can compute the value of each polynomial of degree n

$$a_n \cdot x^n + a_{n-1} \cdot x^{n-1} + \ldots + a_2 \cdot x^2 + a_1 \cdot x + a_0$$

for given values of $a_0, a_1, \ldots, a_n$, and x by n multiplications and n additions only.

We saw that the amount of computer work depends on our skill in the process of the algorithm design. Computing x^{16} provides a still more impressive example than the polynomial evaluation. One can write x^{16} by definition as

$$x \cdot x \cdot x \cdot x \cdot x \cdot x \cdot x \cdot x \cdot x \cdot x \cdot x \cdot x \cdot x \cdot x \cdot x \cdot x \ ,$$

and we see that 15 multiplications have to be executed in order to compute x^{16}. The following representation

$$x^{16} = (((x^2)^2)^2)^2$$

of x^{16} provides us with the following method

$$\begin{array}{cccc} x^2 = x \cdot x \ , & x^4 = x^2 \cdot x^2 \ , & x^8 = x^4 \cdot x^4 \ , & x^{16} = x^8 \cdot x^8 \\ L \leftarrow x \cdot x \ , & L \leftarrow L \cdot L \ , & L \leftarrow L \cdot L \ , & L \leftarrow L \cdot L \end{array}$$

for computing x^{16}. Applying this method, only 4 multiplications suffice to compute x^{16}.

Exercise 5.4 Show how to compute

(a) x^6 using 3 multiplications,
(b) x^{64} using 6 multiplications,
(c) x^{18} using 5 multiplications,
(d) x^{45} using 8 multiplications.

Is it possible to compute x^{45} by fewer than 8 multiplications?

The situation when one has the same complexity for computing any instance of a problem such as the evaluation of a quadratic polynomial is not typical. For the evaluation of quadratic polynomials we got the same complexity for each instance because the problem is simple and our measurement was rough. Our complexity measurement is correct, assuming all concrete input values for a, b, c, and x are numbers that can be saved each in one register of a considered size of 16 or 32 bits. But what happens if the numbers are so large that their representation is several hundreds of bits long? Is it allowed to count the same amount of computer work for an operation over such large numbers as for computing an arithmetic operation over 16-bit integers for which the computer hardware was developed? If in some applications it really happens that one has to compute an arithmetic operation over big numbers, then one is required to write a program that computes this operation by means of the available operations over numbers of a regular size. But we do not want to burden you with this additional technical problem, and therefore, we assume that all forthcoming numbers do not exceed a reasonable size. Because of this assumption, we may measure the time complexity as the number of arithmetic operations, number comparisons, and similar fundamental instructions of a computer or of a programming language.

Using the assumption about restricted sizes of numbers in our computations does not change the fact that it is not typical to have the same amount of computer work for each problem instance. If one has to create a telephone directory by sorting the names for a village with 3000 inhabitants and for a city with two million inhabitants, then, certainly, the amount of work necessary is essentially higher in the second case than in the first case. This is not surprising but this is the kernel of the story. We expect that the complexity of computing a solution has something to do with the **size of the input** (with the amount of input data). For instance, in the case of sorting, one can view the input size as the number of given objects to be sorted (3000 or two million inhabitants). For the problem of computing values of arbitrary polynomials on given arguments, one can take the degree of the polynomial (i.e., the number of coefficients minus one) as the input size. In this case, for the polynomial

$$a_n x^n + a_{n-1} x^{n-1} + \ldots + a_2 x^2 + a_1 x + a_0$$

the input is represented by the following $n + 2$ numbers

$$(a_n, a_{n-1}, a_{n-2}, \ldots, a_2, a_1, a_0, x)\ ,$$

and we order the size n to this problem instance. A naive algorithm A computes the value of a polynomial of degree n as follows:

$$\underbrace{x \cdot x = x^2}_{1^{st}\ multiplication} \qquad \underbrace{x \cdot x^2 = x^3}_{2^{nd}\ multiplication} \qquad \cdots \qquad \underbrace{x \cdot x^{n-1} = x^n}_{(n-1)^{th}\ multiplication}$$

$$\underbrace{a_1 \cdot x}_{n^{th}\ multiplication} \qquad \underbrace{a_2 \cdot x^2}_{(n+1)^{th}\ multiplication} \qquad \cdots \qquad \underbrace{a_n \cdot x^n}_{(2n+1)^{th}\ multiplication}$$

and then summing

$$a_0 + a_1 x + a_2 x^2 + \ldots + a_n x^n$$

by using n additions we get the value of the polynomial for the argument value x.

$$\begin{array}{ccccccc} a_0 & + & a_1 x & + & \cdots & + & a_n x^n \\ & \uparrow & & \uparrow & & \uparrow & \\ & 1^{st}\ addition & & 2^{nd}\ addition & & n^{th}\ addition & \end{array}$$

Hence, the time complexity of the algorithm A is the function

$$\mathrm{Time}_A(n) = \underbrace{2n + 1}_{multiplications} + \underbrace{n}_{additions} = 3n + 1\ .$$

Thus, we imagine that the **time complexity $\mathbf{Time}_A$ of an algorithm A** is

> *a function of the input size that measures the number $Time_A(n)$ of operations that is sufficient and necessary to solve any problem instance of size n using the algorithm A.*

It may happen that different input instances of the same size force different amounts of computer work to be solved. In such a case, $\mathrm{Time}_A(n)$ is considered to be the time complexity of A on the hardest input instance of the size n, i.e., the maximum over the time complexities of the computations of A on all problem instances of size n. As a result of this definition of Time_A, we are on the secure side. In this way one has the guarantee that the algorithm A successfully solves any problem instance of size n by executing at most $\mathrm{Time}_A(n)$ instructions and that there exists a problem instance of size n on which A executes exactly $\mathrm{Time}_A(n)$ instructions.

5.3 Why Is the Complexity Measurement Useful?

As in other scientific disciplines, the discovered knowledge serves to predict the development of various situations and processes of interest. If one was able to determine the time complexity of an algorithm,[1] then one can reliably predict the running time of the algorithm on given problem instances without executing the corresponding computations. Additionally, one can compare the grades (efficiency) of two or more algorithms for the same task by visualizing their complexity functions.

In Fig. 5.1, the curves of two functions, namely $3n - 2$ and n^2, are depicted. The x-axis is used for the input size, and the y-axis corresponds to the time complexity analyzed. We immediately see that the algorithm with time complexity $3n - 2$ is better (more

[1] using the so-called algorithm analysis

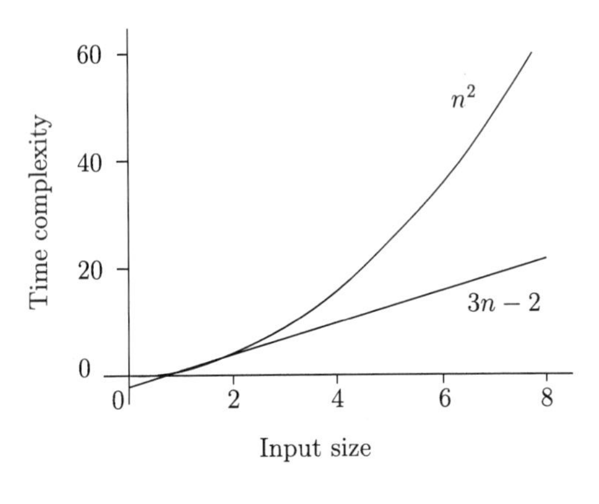

Fig. 5.1

efficient) than the algorithm with complexity n^2 for input sizes greater than 2. One can easily check this by showing that

$$3n - 2 < n^2$$

for all natural numbers greater than 2.

Let us consider another example. Let A and B be two algorithms for a problem U with

$$\mathrm{Time}_A(n) = 2n^2 \text{ and } \mathrm{Time}_B(n) = 40n + 7000 .$$

Since linear functions such as $\mathrm{Time}_B(n)$ increase more slowly than quadratic functions, such as $\mathrm{Time}_A(n)$, the question is, up to what value of the input size is the algorithm B more favorable than the algorithm A? For these special complexity functions we calculate this value here. The question is, for which positive integers does the inequality

$$2n^2 > 40n + 7000$$

hold? This question is equivalent to the following question (we reduce both sides of the inequality by $40n + 7000$): When does

$$2n^2 - 40n - 7000 > 0$$

hold? Dividing both sides of the inequality by 2, we obtain

$$n^2 - 20n - 3500 > 0 . \tag{5.1}$$

Applying the well-known method for solving quadratic equations, we can calculate the solutions of the equation $n^2 - 20n - 3500 = 0$, or simply observe that

$$(n + 50) \cdot (n - 70) = n^2 - 20n - 3500 .$$

Using any of these solution methods we get exactly two solutions -50 and 70 and we see that the inequality 5.1 is true for $n < -50$ and $n > 70$. Because we are interested only in positive integers, we make the following conclusions:

- B is more favorable than A for problem instances of size greater than 70.
- A and B are equally suitable for the input size $n = 70$.
- A is more favorable than B for input sizes between 1 and 69.

Exercise 5.5 Consider three algorithms A, B, and C that solve the same computing problem. Let $\text{Time}_A(n) = n^3/2 + 1$, $\text{Time}_B(n) = n^2 + 7$, and $\text{Time}_C(n) = 5n + 140$ be their corresponding complexities. Try to estimate by calculation and by drawing the function curves which of the three algorithms is the best one for which interval of input sizes.

Another kind of question posed is the following one. A user knows exactly how much time one is allowed to wait for the outcome of the work of an algorithm.

Many interactive applications have a waiting limit below 10 seconds or even much smaller (fractions of a second). If one has a computer executing 10^9 computer instructions per second, then one can perform in 10 seconds computations that consists of at most $10 \cdot 10^9 = 10^{10}$ instructions. Assume we propose to users an algorithm with time complexity $5 \cdot n^3$. They can calculate:

$$\begin{aligned} 5n^3 &< 10^{10} && \mid :5 \\ n^3 &< 2 \cdot 10^9 && \mid \sqrt[3]{\ } \text{ for both sides} \\ n &\leq 1250 && . \end{aligned}$$

Following these calculations, users know that they can apply the algorithm for input instances of size 1250 and they will successfully obtain the results in time. Usually, users know exactly the typical sizes of forthcoming input instances, and so they can immediately decide whether they want to use this algorithm or they ask for a faster one.

Assume we have an optimization problem that has a huge number of solutions, and our task is to find the best one with respect to a well-defined criterion. This task is very important because it is related to a big investment such as extending the traffic (railway or highway) network in a country or finding a placement of senders and receivers for a wireless network. Because the differences in costs between distinct solutions may be of the order of many millions, there is a strong willingness to take enough time to find a really good solution and to invest money to pay for expensive fast computer systems and professional algorithmicists searching for good solutions. Assume the project management operates at the limits of physically doable and poses the time limit by accepting computations executing up to 10^{16} operations. We can visualize this limit as a line $y = 10^{16}$ parallel to the x-axis as depicted in Fig. 5.2. One can see the limit on applying an algorithm A by looking at the intersection of the line $y = 10^{16}$ and the curve of the function $\mathrm{Time}_A(n)$. The x coordinate n_A of this intersection is the upper bound on the size of input instances for which one can guarantee a successful computation in time. Hence, we can look at our specific large problem instance (better to say at its size) and we see immediately whether the algorithm is suitable for us or not.

To realize the importance of efficiency, let us analyze this situation for a few time complexities. Let $\mathrm{Time}_A(n) = 3n - 2$. Then we calculate

$$\begin{aligned} 3n - 2 &\leq 10^{16} && \mid +2 \text{ to both sides} \\ 3n &\leq 10^{16} + 2 && \mid \text{divide both sides by } 3 \\ n &\leq \tfrac{1}{3}(10^{16} + 2) = n_A\,. \end{aligned}$$

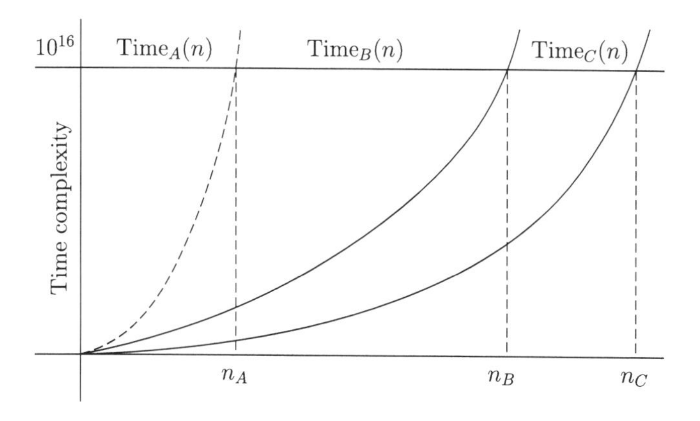

Fig. 5.2

We see that such a huge input size will never be considered, and so the algorithm A is always applicable. For an algorithm B with $\mathrm{Time}_B(n) = n^4$ we obtain

$$\begin{aligned} n^4 &\leq 10^{16} && \mid \sqrt[4]{\ } \text{ of both sides} \\ n &\leq (10^{16})^{1/4} = 10^4 = 10,000 \ . \end{aligned}$$

Hence, the algorithm B is up to $n_B = 10000$ practicable. Because most typical applications consider essentially smaller input sizes, one still has to view B as a good, efficient algorithm.

Assume now that $\mathrm{Time}_C(n) = 10^n$. Then, the analysis looks as follows:

$$\begin{aligned} 10^n &\leq 10^{16} \mid \log_{10} \text{ of both sides} \\ n &\leq 16 \ . \end{aligned}$$

Hence, the situation is not good. In spite of a huge time limit of 10^{16} instructions, we are able to solve small problem instances only. If one considers the exponential function $f(n) = 2^n$, then one observes that

$$2^{n+1} = 2 \cdot 2^n \ ,$$

i.e., that +1 in the input size doubles the amount of computer work. Hence, one can conclude that algorithms with an exponential time complexity have very restricted applicability.

Exercise 5.6 Assume a computer scientist improves an algorithm C with complexity $\mathrm{Time}_C(n) = 10^n$ to an algorithm D with $\mathrm{Time}_D(n) = 4 \cdot (1.2)^n$. How much does the situation improve in this way? How large problem instances can be solved in the limit of 10^{16} operations using the new algorithm D?

Exercise 5.7 Assume one has a computer, executing 10^9 instructions per second. The age of the Universe is smaller than 10^{18} seconds. We are willing to wait for 10^{18} seconds. Problem instances of which size can be solved using an algorithm A if its time complexity is

(i) $\mathrm{Time}_A(n) = 10 \cdot n^2$?
(ii) $\mathrm{Time}_A(n) = 50 \cdot n^3$?
(iii) $\mathrm{Time}_A(n) = 2^n$?
(iv)* $\mathrm{Time}_A(n) = n! = n \cdot n(n-1) \cdot (n-2) \cdot \ldots \cdot 2 \cdot 1$? (challenge)

5.4 Limits of Tractability

Above, we saw that the time complexity of an algorithm decides its applicability. But our aim is more demanding than analyzing the complexity of designed algorithms. We want to measure the hardness of algorithmic problems in order to be able to decide if they are **tractable** (practically solvable) or not. At first glance, the way from the measurement of complexity of algorithms to the measurement of the complexity of problems looks simple. One could propose the following definition:

The complexity of a problem U is the complexity of the best (time-optimal) algorithm for U.

Though this specification of problem hardness seems to be reasonable, one cannot use it in general for any problem. Scientists showed that there are problems that do not possess any algorithm that could be called the best one for them. For such a special task U one can essentially[2] improve each algorithm solving U. In this way, improving algorithms for U is a never-ending story.

[2] for infinitely many input sizes

Since we cannot in general identify the complexity of any problem U with one of the best algorithms solving U, we speak about lower and upper bounds on the complexity of a problem.

Definition 5.1 Let U be a problem and let A be an algorithm that solves U. Then we say that the time complexity $\mathrm{Time}_A(n)$ of the algorithm A is an **upper bound of the time complexity of U.** For a function f, we say that $f(n)$ is a **lower bound of the time complexity of U** if there does not exist any algorithm B for U with

$$\mathrm{Time}_B(n) \leq f(n)$$

for almost all[3] n.

This definition of problem hardness is sufficient for discovering several important facts. For instance, there exist arbitrary hard algorithmically solvable problems. To say it more precisely, for any quickly growing function such as $2^n, n!$ or 2^{2^n}, one can find problems that are solvable within this complexity but unsolvable within any smaller complexity. Most such hard (intractable) problems were constructed using a more elaborate version of the diagonalization method and so they are more or less artificial.

Estimating the complexity of concrete, real problems occurring in practice is more complex. We know a few thousand problems with an exponential upper bound about 2^n on their complexity (because the fastest algorithms designed for those problems require such a huge amount of computer work), for which we miss a lower bound growing faster than the linear functions. This means that we are unable even to prove small lower bounds as $n \cdot \log n$ or n^2 for problems whose complexity is probably exponential. In other words, we have many problems with a huge gap between linear lower bounds $c \cdot n$ and exponential upper bounds such as 2^n, and we are unable to estimate the complexity of these problems more precisely. The 40 years of unsuccessful attempts can be explained by the hardness of proving lower bounds on the complexity of concrete problems.

[3] for all but finitely many

Currently, one considers proving lower bounds on complexity and so proving nonexistence of efficient algorithms for concrete problems as the hardest core of computer science. There even exist proofs showing that current mathematical methods are not sufficient for proving the aimed lower bounds on the complexity of concrete problems. This means that some development of proof methods in mathematics (i.e., of mathematical arguments) is necessary in order to be able to derive higher lower bounds on the complexity of concrete algorithmic tasks.

Investigation of the complexity of algorithms and problems posed a new question that became the key question of algorithmics:

> *Which algorithmically solvable problems are practically solvable (tractable)?*
> *Where is the limit of practical solvability (tractability)?*

Following our investigation of complexity functions in Section 5.3 and the content of Table 5.1, we see that algorithms with an exponential complexity cannot be viewed as useful in practice. In Table 5.1 one presents the number of instructions to be executed for 5 complexity functions $10n, 2n^2, n^3, 2n$, and 2^n and for four input sizes $10, 50, 100$, and 200.

n / $f(n)$	10	50	100	300
$10n$	100	500	1000	3000
$2n^2$	200	5,000	20,000	180,000
n^3	1000	125,000	1,000,000	27,000,000
2^n	1024	16 digits	31 digits	91 digits
$n!$	$\approx 3.6 \cdot 10^6$	65 digits	158 digits	615 digits

Table 5.1

If the number of operations is too large, we present the number of its digits instead of the number itself. We immediately see that algorithms with exponential complexity functions such as 2^n and $n!$ are not applicable already for problem instances of a size above 50.

After some years of discussions, computer scientists agreed on the following definition of the tractability of computing tasks:

> *An algorithm A with $Time_A(n) \leq c \cdot n^d$ for some constants c and d (for concrete numbers c and d) is called a polynomial algorithm.*
> *Each problem that can be solved by a polynomial algorithm is called* ***tractable (practically tractable). P*** *denotes the class of all tractable decision problems.*

It was not so easy to accept this definition of tractability. Today, we do not consider the limits of tractability given by this definition as sharp limits between tractable problems and intractable ones, but as a first attempt to approximate the limits of algorithmically doable. There were two reasons for accepting this definition of tractability as a working base for classifying algorithmic problems into practically solvable and practically unsolvable.

1. Practical Reason

The main reason is based on experience with algorithm design. The fact that algorithms with exponential complexity are not practical was obvious to everybody. Analysis and experience proved the applicability of algorithms with time complexity up to n^3, and in some applications even up to n^6. But a polynomial algorithm with complexity about n^{100} is for realistic sizes of input data even less applicable than algorithms with complexity about 2^n, because $n^{100} > 2^n$ for most reasonable input sizes n. Nevertheless, experience has proved the reasonability of considering polynomial-time computations to be tractable. In almost all cases, once a polynomial algorithm has been found for an algorithmic problem that formerly appeared to be hard, some key insight into the problem has been gained. With this new knowledge, new polynomial-time algorithms with a lower degree than the former ones were designed. Currently, there are only a few known exceptions of nontrivial problems for which the best polynomial-time algorithm is not of practical utility. Therefore, from a practical point of view, one does not consider the class P to be too large, and typically the problems from P are viewed as tractable.

2. **Theoretical Reason**

Any definition of an important problem class has to be robust in the sense that the defined class is invariant with respect to all reasonable models of computation. We cannot allow that a problem is tractable for one programming language (e.g., Java) but not for another computing model or for another programming language. Such a situation would arise if one defines the set of problems solvable within time complexity $c \cdot n^6$ as the class of practically solvable problems.[4] The crucial fact is that the class P is robust with respect to all reasonable computing models. The class of problems solvable in polynomial time is the same for all reasonable models of computation. The consequence is that proving the membership or the non-membership of a problem into the class P has general validity and can be reliably used for classifying problems into practically solvable and practically unsolvable.

5.5 How Do We Recognize a Hard Problem?

The main task of complexity theory is to classify concrete computing problems with respect to their degree of hardness measured by computational complexity. Designing algorithms, one obtains upper bounds on the complexity of problems. The trouble is that we are unable to derive lower bounds on the complexity of concrete problems. How can we then classify them? In fact, we cannot. Therefore, we do what scientists and also non-scientists with well-developed human intellect in such situations usually do. Instead of performing countless attempts at running headfirst against a wall in order to exactly estimate the complexity of a problem, we

[4] For instance, some theoretical computer models require a quadratic increase of time complexity in order to be able to simulate programs written in some programming languages.

are satisfied with plausible complexity estimations, whose absolute reliability we are unable to prove.

What does it mean to argue "plausibly" (but not reliably) that there is no polynomial algorithm for a problem? Currently, we know more than 4000 interesting computing tasks for which, in spite of huge effort, no efficient algorithm has been found. But this negative experience does not suffice for claiming that all these problems are computationally hard. This is not the way out. For problems such as primality testing and linear programming, mathematicians and computer scientists tried to find polynomial algorithms for many years[5] without any success. And then we celebrated the discoveries providing polynomial algorithms for these problems. In spite of the fact most of us believed in the existence of efficient algorithms for these problems, this experience unmistakably shows the danger of making a mistake by claiming a problem is hard (not practically solvable) only because we were not successful in solving it in an efficient way.

S.A. Cook and L.A. Levin developed an idea for how to join all these negative experiences into a plausible assumption. They proposed the following definition of problem hardness:

> *A problem U is hard (not in P or not practically solvable), if the existence of a polynomial algorithm for U implies also the existence of polynomial algorithms for several thousand other problems that are considered to be hard (i.e., for which one is unable to find efficient algorithms).*

We describe this idea transparently in Fig. 5.3. The class P is on the left side of the picture in Fig. 5.3. On the right side, there is a class of several thousand problems, for which nobody was able to design any efficient algorithm. Now assume one could find an efficient algorithm for a problem U from this class, and the existence of a polynomial algorithm for U implies the existence of efficient algorithms for all these many problems in the class of the right side.

[5] In the case of primality testing, even more than 2000 years.

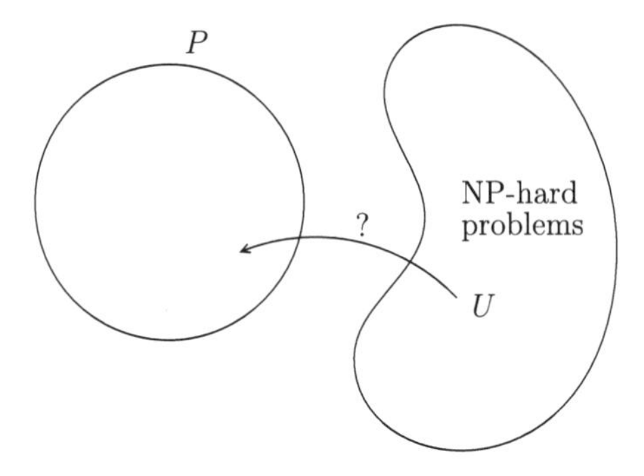

Fig. 5.3

Then the hardness of U seems to be more plausible. Currently, almost nobody believes that all these problems can be efficiently solved, in spite of the fact we were unable to solve any of them efficiently. We are unable to consider ourselves to be so foolish that after numerous unsuccessful attempts to solve a large variety of different problems all these problems admit efficient algorithms solving them.

Problems where efficient solvability directly implies efficient solvability of many other problems that are currently viewed as being hard are called NP-hard problems. The main question is now the following one:

How to show that a problem is NP-hard?

Once again, the method of reduction helps us solve this problem. We applied the reduction method in order to show that algorithmic solvability of a problem implies algorithmic solvability of another problem. Now one needs to exchange the predicate "algorithmic" with the predicate "efficiently". One can do this using the framework of efficient reductions. As depicted in Fig. 5.4, one efficiently reduces a problem U_1 to a problem U_2 by designing an efficient algorithm R that transforms any instance I of U_1 into an equivalent instance $R(I)$ of U_2.

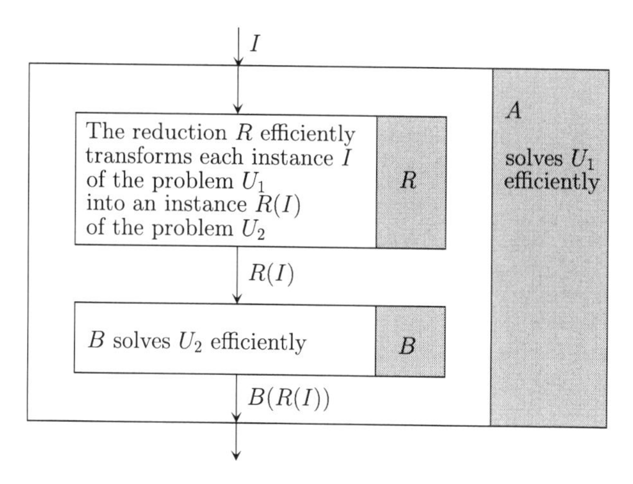

Fig. 5.4

The term "equivalent" means that the solutions for the instance I of the problem U_1 and the solution for the instance $R(I)$ of the problem U_2 are the same. In this way, the output $B(R(I))$ of the computation of the algorithm B on $R(I)$ can be directly taken as the output (solution) for I. In fact, we presented an efficient reduction in Fig. 4.8 in Chapter 4. There we designed an efficient reduction of the problem of solving general quadratic equations to the problem of solving quadratic equations in the p, q-form. The consequence is that efficient solvability of quadratic equations in p, q-form implies efficient solvability of general quadratic equations.

We say that the algorithm R in Fig. 5.4 is a **polynomial reduction from U_1 to U_2** if R is a polynomial algorithm with the property:

> *The solution for the instance I of U_1 is identical to the solution for the instance $R(I)$ of U_2 for all instances of the problem U_1.*

If U_1 and U_2 are decision problems, this means that either "YES" is the right answer for both I and $R(I)$ or "NO" is the correct output for both I and $R(I)$. In such a case we say that the instance $R(I)$ of U_2 is **equivalent** to the instance I of U_1.

If there exists a polynomial reduction R from U_1 to U_2, then we say that

U_1 is polynomial-time reducible to U_2,

and write

$$\boldsymbol{U_1 \leq_{pol} U_2.}$$

Similarly to the case of the general reduction, $U_1 \leq_{pol} U_2$ means that U_2 is not easier than U_1 with respect to polynomial-time (efficient) solvability, i.e., that either U_1 and U_2 are equally hard (both are efficiently solvable or neither of them is efficiently solvable), or U_1 is efficiently solvable but U_2 is not. The only excluded situation is that U_2 is efficiently solvable and U_1 is not.

Exercise 5.8 Show that the problem of calculating the height of an isosceles triangle with known side lengths is polynomial-time reducible to the problem of computing an unknown length of a side of a right-angled triangle.

Exercise 5.9 Let U_2 be the problem of solving linear equations of the form $a + bx = 0$. Let U_1 be the problem of solving linear equations of the general form $a + bx = c + dx$. Prove that $U_1 \leq_{pol} U_2$.

The notion of polynomial reducibility very fast became a successful instrument for the classification of problems with respect to tractability. Currently we know a few thousand problems that are polynomial-time reducible to each other. And $U_1 \leq_{pol} U_2$ together with $U_2 \leq_{pol} U_1$ means that U_1 and U_2 are equally hard in the sense that either both are solvable in polynomial-time or neither of them is efficiently solvable. Therefore, we know a few thousand problems that are either all efficiently solvable or none of them is tractable. These problems are called NP-hard.[6]

The reduction examples presented so far do not really show the applicability of the reduction method in complexity theory because all problems considered there are simply solvable tasks. Therefore,

[6] We do not give the formal definition of NP-hardness because it works with concepts we did not introduce here.

our next step is to show a reduction between two NP-hard problems.

As our first problem U_1, we consider the supervision problem that is called the vertex cover problem and is denotated by VC in algorithmics. Any problem instance is described as a network of roads (streets) with n crossroads (called vertices in the computer science terminology) and s roads between the crossroads for some natural numbers r and s (Fig. 5.5). The end of the cul-de-sac is also considered a vertex (crossroad).

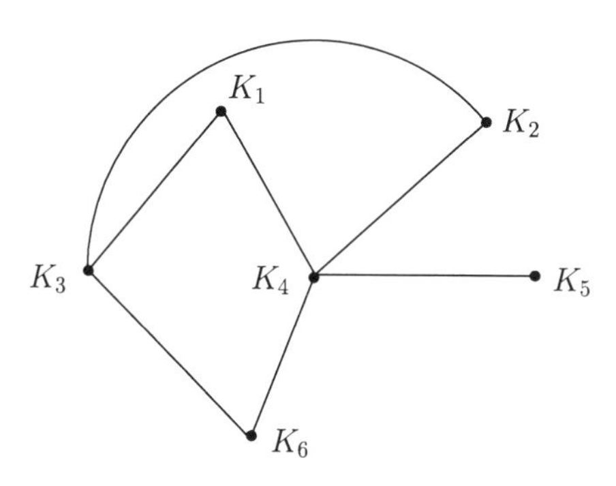

Fig. 5.5

The crossroads are depicted as small black points in Fig. 5.5 and are denoted by K_1, K_2, K_3, K_4, K_5, and K_6. The lines between the points correspond to the streets. In Fig. 5.5 we have 7 streets. One can also name all the streets. We call the street between K_1 and K_3 simply $Str(K_1, K_3)$. Since we do not consider any direction for the streets, the denotations $Str(K_1, K_3)$ and $Str(K_3, K_1)$ represent the same street. One is allowed to post a supervisor at any crossroad. One assumes that the supervisor can supervise all the streets outgoing from his crossroad over the whole length up to the neighboring crossroads. In this way, a supervisor at crossroad K_3 can observe the streets $Str(K_3, K_1)$, $Str(K_3, K_2)$, and $Str(K_3, K_6)$. In Fig. 5.6 we see all the streets supervised from crossroad K_3 as dashed lines.

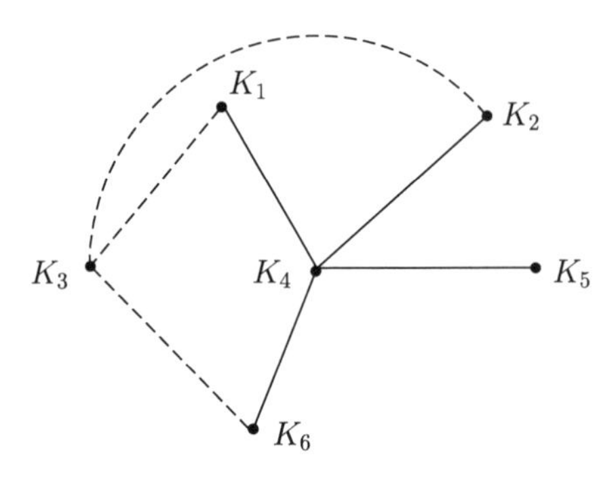

Fig. 5.6

A further part of the problem instance is a natural number m. The question is whether m supervisors suffice to supervise all streets. In other words, is it possible to assign m supervisors to m crossroads in such a way that each street is supervised by at least one supervisor? For the network of roads in Fig. 5.5 and $m = 2$, there is a solution. If one supervisor is at K_4, then he supervises the four streets $Str(K_4, K_1)$, $Str(K_4, K_2)$, $Str(K_4, K_5)$, and $Str(K_4, K_6)$. The second supervisor at K_3 can supervise the other streets.

Exercise 5.10 Assume that we are not allowed to position any supervisor at the crossroad K_4 in the street network in Fig. 5.5. How many supervisors do we need in this case?

Exercise 5.11 Consider the road network in Fig. 5.7. Do three supervisors suffice for supervising all streets?

Exercise 5.12 Draw a new street $Str(K_1, K_2)$ between K_1 and K_2 in Fig. 5.7. How many supervisors are necessary in order to be able to supervise all streets of this road network?

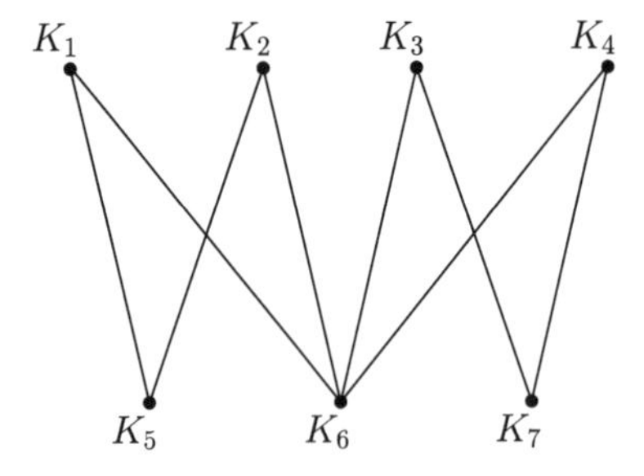

Fig. 5.7

Exercise 5.13 Add two new streets $Str(K_5, K_6)$ and $Str(K_2, K_5)$ to the road network in Fig. 5.5. Do three supervisors suffice in order to supervise the network?

As the second problem U_2 we consider the problem **LIN(0,1)**, where one has to estimate whether a system of linear inequalities has a Boolean solution. For instance,

$$x_1 + 2x_2 - 3x_3 + 7x_4 \geq 3$$

is a linear inequality with 4 unknowns x_1, x_2, x_3, and x_4. We say that this inequality is over Boolean variables (unknowns) if one allows only Boolean values 0 and 1 for the unknowns x_1, x_2, x_3, and x_4. An instance of **LIN(0,1)** is the following system of linear inequalities:

$$\begin{aligned} x_1 + 2x_2 + x_3 + x_4 &\geq 3 \\ x_1 + x_4 &\geq 0 \\ 2x_1 + x_2 - x_3 &\geq 1. \end{aligned}$$

This is a system of three linear inequalities and four unknowns. The task is to decide whether one can choose values 0 and 1 for the unknowns in such a way that all inequalities are at once valid. For instance, we can take the value 1 for x_1 and x_2 ($x_1 = x_2 = 1$) and the value 0 for x_3 and x_4 ($x_3 = x_4 = 0$). We see that

$$\begin{aligned} x_1 + 2x_2 + x_3 + x_4 &= 1 + 2 \cdot 1 + 0 + 0 = 3 \geq 3, \\ x_1 + x_4 &= 1 + 0 = 1 \geq 0, \\ 2x_1 + x_2 - x_3 &= 2 \cdot 1 + 1 - 0 = 3 \geq 1, \end{aligned}$$

and so all inequalities are satisfied.

Exercise 5.14 Find other Boolean values for the unknowns x_1, x_2, x_3, and x_4 of the considered system of linear inequalities that also satisfy all three inequalities.

Exercise 5.15 Does the following system of linear inequalities have a Boolean solution?

$$\begin{aligned} x_1 + x_2 - 3x_3 &\geq 2 \\ x_1 - 2x_2 - x_4 &\geq 0 \\ x_1 + x_3 + x_4 &\geq 2 \end{aligned}$$

We show now that

$$\mathrm{VC} \leq_{pol} \mathrm{LIN}(0,1)$$

i.e., that the supervision problem VC is polynomial-time reducible to the problem of linear inequalities with Boolean variables.

To do this, we have to efficiently construct an equivalent instance of LIN(0,1) to any instance of VC. We explain the idea of the reduction by creating an instance of LIN(0,1) to the road network N in Fig. 5.5. Let $(N, 3)$ be the problem instance of VC. To solve it, one has to decide whether 3 supervisors are sufficient for supervising the network N. To pose this question in terms of linear programming, we take 6 Boolean variables x_1, x_2, x_3, x_4, x_5, and x_6. The variable x_i is assigned to the crossroad K_i, and its meaning is the following one:

$x_i = 1$, a supervisor is placed in K_i,
$x_i = 0$, no supervisor is in K_i.

For instance, the value assignment $x_1 = 1, x_2 = 0, x_3 = 1, x_4 = 0, x_5 = 1$, and $x_6 = 0$ describes the situation where one has 3 supervisors placed at the crossroads K_1, K_3, and K_5, and there is no supervisor at K_2, K_4, and K_6.

First, we use the linear inequality

$$x_1 + x_2 + x_3 + x_4 + x_5 + x_6 \leq 3$$

to express the constraint that at most 3 supervisors can be used. To supervise the network N, each street of N has to be supervised by a supervisor. This means that, for each street $Str(K_i, K_j)$ between the crossroads K_i and K_j, at least one supervisor has to be placed at one of the crossroads K_i and K_j. For the street $Str(K_1, K_4)$, at least one supervisor must be at K_1 or K_4. To express this constraint, we use the linear inequality

$$x_1 + x_4 \geq 1 \ .$$

This inequality is satisfied if $x_1 = 1$ or $x_4 = 1$, and this is exactly what we need. If one takes all 7 streets (K_1, K_3), (K_1, K_4),

(K_2, K_3), (K_2, K_4), (K_3, K_6), (K_4, K_5), and (K_4, K_6) of the network N into account, then one gets the following system $L1$ of 8 linear inequalities:

$$\begin{aligned} x_1 + x_2 + x_3 + x_4 + x_5 + x_6 &\leq 3 \ \{\text{at most 3 supervisors}\} \\ x_1 + x_3 &\geq 1 \ \{Str(K_1, K_3) \text{ is supervised}\} \\ x_1 + x_4 &\geq 1 \ \{Str(K_1, K_4) \text{ is supervised}\} \\ x_2 + x_3 &\geq 1 \ \{Str(K_2, K_3) \text{ is supervised}\} \\ x_2 + x_4 &\geq 1 \ \{Str(K_2, K_4) \text{ is supervised}\} \\ x_3 + x_6 &\geq 1 \ \{Str(K_3, K_6) \text{ is supervised}\} \\ x_4 + x_5 &\geq 1 \ \{Str(K_4, K_5) \text{ is supervised}\} \\ x_4 + x_6 &\geq 1 \ \{Str(K_4, K_6) \text{ is supervised}\} \end{aligned}$$

Now, the following is true. The system $L1$ has a solution if and only if the instance $(N, 3)$ of VC has a solution (i.e., if and only if three supervisors can supervise N). The claim can even be formulated in a stronger way. Each solution of $L1$ provides a solution for $(N, 3)$. For instance,

$$x_1 = 0, x_2 = 1, x_3 = 1, x_4 = 1, x_5 = 0, x_6 = 0$$

is a solution for $L1$. We have only three supervisors in K_2, K_3, and K_4 and so the first inequality is satisfied. The value assignment $x_2 = 1$ ensures that the fourth inequality and the fifth inequality are satisfied. The inequalities 2, 4, and 6 are satisfied due to $x_3 = 1$. The assignment $x_4 = 1$ guarantees the fulfilment of inequalities 3, 5, 7, and 8. Hence, all 8 inequalities are satisfied. We immediately see that three supervisors in K_2, K_3, and K_4 supervise the network N.

Exercise 5.16

(a) Find all solutions for the system $L1$ (i.e., all assignments of values 0 and 1 to the variables $x_1, x_2, \ldots, x_6$ that satisfy all inequalities of $L1$) and estimate the corresponding solutions for $(N, 3)$.
(b) Does there exist a solution for $L1$ with $x_4 = 0$ (i.e., if no supervisor is placed in K_4)? Argue for the correctness of your answer.
(c) Extend the road network N by the street $Str(K_3, K_4)$ to a network N'. Does $(N', 2)$ have a solution?

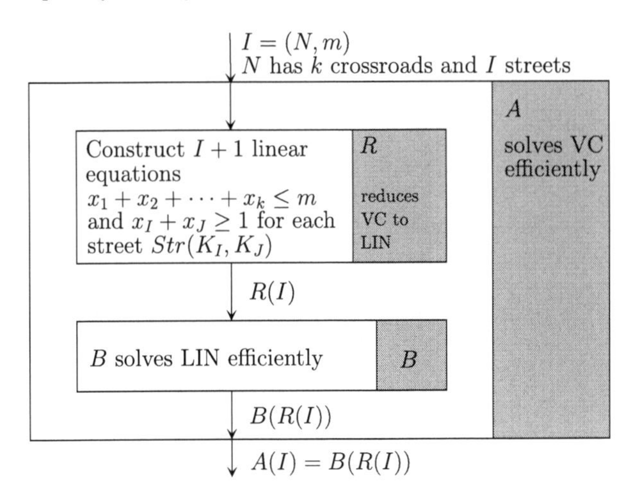

Fig. 5.8

A general description of the polynomial reduction of VC to LIN(0,1) is given in Fig. 5.8.

We see that the reduction R can be easily performed by a program. The efficiency of R ensures that the existence of an efficient algorithm B for LIN(0,1) implies the existence of an efficient algorithm for VC.

Exercise 5.17 Consider the instance $(N, 3)$ of VC for the network N in Fig. 5.7. Use the description of the reduction in Fig. 5.8 to construct the corresponding system of linear inequalities. Find all solutions of the constructed system of linear inequalities and the corresponding solutions for $(N, 3)$.

Exercise 5.18 Extend the network N in Fig. 5.7 to a network N' by adding two new streets $Str(K_2, K_3)$ and $Str(K_5, K_6)$. What about the system of linear inequalities for $(N', 4)$? Does it have a solution?

We proved the relation VC $\leq_{pol}$ LIN$(0, 1)$. One can show also LIN$(0, 1)$ $\leq_{pol}$ VC. But this reduction is too technical, and therefore we omit its presentation here.

Hence, the problems VC and Lin(0,1) are equally hard in the sense that the existence of a polynomial algorithm for one of the problems directly implies the existence of an algorithm for the other problem. Currently, we know more than 4000 equally hard problems, and for none of them do we have an efficient algorithm. Therefore, we believe that these problems are really hard, and so we use the polynomial reduction to classify problems into easy ones solvable in polynomial-time and hard ones that are probably not solvable in polynomial-time. For a new problem U, it is sufficient to show

$$U' \leq_{pol} U$$

for a problem U' from this class of hard problems and we are allowed to view U as a hard problem. The fact $U' \leq_{pol} U$ directly implies that the existence of an efficient algorithm for U provides the existence of efficient algorithms for thousands of problems considered to be hard.

Computer scientists developed this approach for arguing for the hardness of concrete algorithmic tasks because they did not master the development of techniques for proving lower bounds on complexity. One cannot exclude the possibility that someone may devise a brilliant idea providing efficient algorithms for all NP-hard problems. Are we allowed to believe in the hardness of NP-hard problems only because the experience of researchers in algorithmics supports this opinion? One can give a pragmatic answer to this question. Even if NP-hard problems are efficiently solvable, currently they are not efficiently solvable, and this state is not going to change up to the moment one discovers an efficient algorithm for an NP-hard problem. What are we trying to say? Independently of whether there exist efficient algorithms for NP-hard problems or not, currently we do not know any efficient way for solving NP-hard problems, and so the NP-hard problems have to be considered to be hard for us. Researchers working in complexity theory have additional serious reasons to believe in the real hardness of NP-hard problems. They proved that a consequence of efficient solvability of NP-hard problems would mean that, for a large class of claims, to find proofs of the truthfulness of these claims is algorithmically as hard as verifying the correctness of given proofs. Nobody wants to believe that creating proofs (which is one of the hardest intellectual jobs of mathematicians) is not harder than checking the correctness of already created proofs.

Now, the reader may pose the following question. If computer scientists are convinced of the fact that there are no polynomial algorithms for NP-hard tasks, why do they not postulate this fact as a new axiom of computer science? The answer is: "They are not allowed." One may postulate only those assertions and definitions as axioms that are not provable. As we explained in Chapter 1, there is no way to prove the correctness of axioms. If they are wrong, there is only a way to disprove them. If one develops successful methods for proving lower bounds on the complexity of concrete problems, one could be able to prove the hardness of the NP-hard problems. Therefore, we did not blindly believe the real hardness of the NP-hard problems and invest a lot of effort in justifying it. This goal is not only one of the key research problems of computer science, mathematicians consider this question to be one of the most challenging problems of mathematics.

5.6 Help, I Have a Hard Problem ...

One is not allowed to underestimate this call for help. The fastest algorithms for NP-hard problems require for realistic input sizes more work than one can perform in the Universe. Is it not a sufficient reason for a call for help? Especially if the solutions to given problems are related to security or to a big investment. Another question about the reasonability is the following one. Is there somebody who can help? The answer is "Yes". Specialists in algorithmics can help. They are real artists and magicians in searching for solutions to problems. If they study a hard problem, they apply a strategy based on the fact that many hard problems are very sensitive (unstable) in the following sense: A small change in the problem specification or in the constraints describing feasible solutions can cause a huge change in the amount of computer work sufficient and necessary for calculating solutions. Usually, weakening of one of the constraints on feasible solutions can save a lot of computer work. Often this constraint reduction is acceptable for the application considered, and it reduces a billion years of computer work to a matter of a few seconds on a common PC. Similar effects are possible if one forces the algorithm to find a near-optimal solution instead of an optimal one. For instance, "near-optimal" could mean that a measurable solution quality of the solution computed differs at most by 1% from the quality of the optimal solutions. Algorithms computing near-optimal solutions are called approximation algorithms. Another approach is provided by randomized algorithms, which randomly choose a strategy for searching for a solution. If the randomly chosen strategy is good for the input instance, then the algorithm efficiently computes the correct solution. If the randomly chosen strategy is not suitable for the given problem instance, then the computation of results may be very long or the computed result may even be wrong. In many applications we are satisfied with randomized algorithms if they efficiently compute the right solution with high probability. To understand this properly, one saves a lot of computer work due to the reduction of the requirement to always compute the correct result to the requirement to compute the correct result with high probability. There are also other approaches for attacking hard problems, and one can combine them. To discover the weak points of hard problems is the art mastered by algorithm designers. The job of algorithm designers is to pay for a jump from an intractable amount of computer work to a realistic amount of work with the minimum possible reduction in our requirements. In some cases, it is possible to pay so little for a large quantitative jump in the amount of computer work that one speaks about miracles. An almost unbelievable miracle of this kind is presented in the next chapter about randomized algorithms.

Here, we present a simple example of saving computer work by computing a relatively "good" solution instead of an optimal one. In Section 5.5 we introduced the problem VC as a decision problem. Its optimization version MIN-VC asks us to find a solution with the minimal number of supervisors sufficient for supervising a given network of roads. For the network in Fig. 5.5, two supervisors can supervise the network and one cannot, i.e., the number 2 is the cost of the optimal solution.

Exercise 5.19 What is the minimal number of supervisors for the roads network in Fig. 5.7? Argue for the correctness of your answer!

To keep our material accessible to non-specialists, we omit presenting spectacular results that are connected with too complex, technical arguments. Rather, we transparently outline the power of the approximation concept by designing an efficient algorithm for computing solutions for the VC problem, whose number of supervisors is at most twice as large as the optimal number. Although this approximation may look too weak, it is far from being trivial for problem instances with 10,000 crossroads. For large networks, we are unable to provide ad hoc solutions with a better quality guarantee.

The idea of searching for a solution with at most double the number of observers is based on the fact that each street $Str(K_1, K_2)$ between two crossroads K_1 and K_2 can be supervised only from K_1 or from K_2. Hence, a supervisor must be placed in at least one of the two crossroads K_1 and K_2. Following this idea one can develop the following algorithm that surely finds a feasible[7] solution.

Algorithm: App-VC

Input: A network N

Procedure:

1. *Choose an arbitrary street* $Str(K_1, K_2)$. *Place two supervisors at both crossroads* K_1 *and* K_2. *Reduce the given network* N *to a network* N' *by removing all streets that are supervised from the crossroads* K_1 *and* K_2 *(in this way, one deletes all streets outgoing from* K_1 *and* K_2*, and so the street* $Str(K_1, K_2)$ *too).*
2. *Use the procedure described in 1 for the network* N'.
3. *Apply this procedure until all streets are removed (i.e., until all streets are supervised).*

[7] A solution is called feasible if it ensures the supervision of all roads of the given network.

Let us execute the work of algorithm App-VC on the networks in Fig. 5.9(a) with 8 crossroads a, b, c, d, e, f, g, and h.

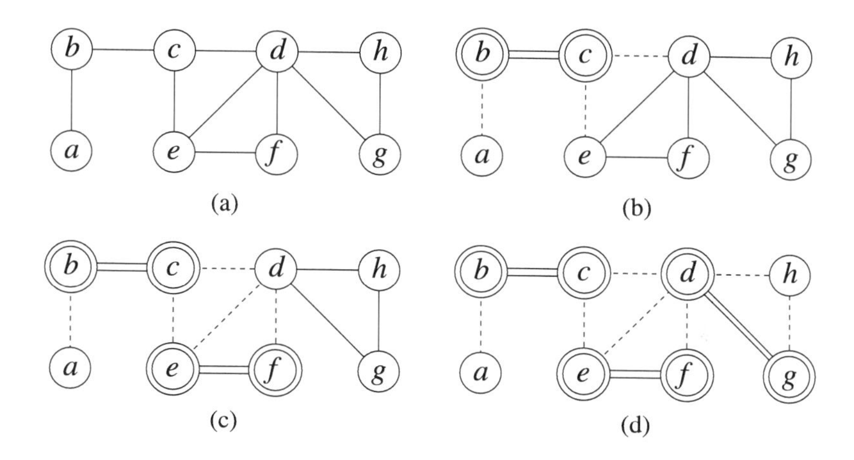

Fig. 5.9

Assume that App-VC chooses the street $Str(b, c)$. This choice is visualized by the double line in Fig. 5.9(b). Now, two supervisors are placed at b and c, and all streets outgoing from b and c are deleted from the network. The deleted streets, except $Str(b, c)$, are marked by the dashed lines in Fig. 5.9(b). Assume the next choice of App-VC is the street $Str(e, f)$ (Fig. 5.9(c)). The procedure is depicted in Fig. 5.9(c). Two observers are positioned at the crossroads e and f, and the streets $Str(e, f), Str(e, d)$ and $Str(f, d)$ are marked as supervised. After this, only three streets, $Str(d, g), Str(d, h)$ and $Str(h, g)$, remained unsupervised in the network. If App-VC chooses street $Str(d, g)$, two new supervisors are placed at the crossroads d and g and the remaining three streets are supervised. This application of App-VC leads to a solution with 6 observers placed at the crossroads b, c, e, f, d, and g.

We observe that the results computed by App-VC may differ and what result is computed depends on the random choice of still unsupervised streets in the first step of the procedure.

Exercise 5.20 How many supervisors does one need in order to supervise the road network in Fig. 5.9(a)? Find such an unfortunate choice of edges by App-VC that at the end the supervisors are placed at all 8 crossroads of the network. Does there exist a random choice of unsupervised edges that results in fewer than 6 supervisors?

Exercise 5.21 Apply the algorithm App-VC in order to place supervisors in networks in Fig. 5.5 and Fig. 5.7.

Finally, we argue why the solutions produced by App-VC cannot place more supervisors than twice the optimum. Let us consider the chosen streets that are marked by double lines in Fig. 5.9. None of these streets meet at a crossroad. This is the consequence of the fact that, after taking the street $Str(K_1, K_2)$, all streets outgoing from K_1 and K_2 were deleted for all remaining choices. Hence, no street chosen later can go to K_1 and K_2. If one takes a global look at a network with marked chosen streets (Fig. 5.10), one sees that these are isolated in the sense that no pair of them meets at a crossroad.

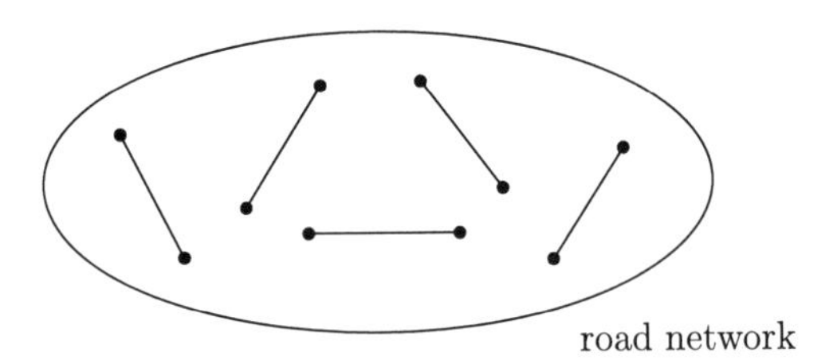

Fig. 5.10

A feasible solution of our VC problem requires that all streets are supervised. Consequently, all chosen (isolated) streets as a special subset of all streets have to be supervised, too. Since the chosen streets are isolated, one supervisor at a crossroad can supervise at most one of these streets. Therefore, any feasible solution has at least as many supervisors as the number of isolated streets. The solution of App-VC places supervisors at all end points (crossroads) of the isolated streets. In this way, the number of observers in the solution computed by App-VC is exactly twice the number of chosen edges. We see that App-VC cannot compute any solution that would require more than double the optimal number of supervisors.

Certainly, the guaranteed quality of App-VC is not impressive. But there are several complex examples of efficient algorithms for NP-hard problems which compute solutions whose quality does not differ from the quality of optimal solutions by more than 1‰. Providing the guarantee for such high-quality solutions can be viewed as a miracle of algorithmics.

5.7 Summary

After the notions of algorithms and programs, the notion of complexity provides the most fundamental concept of computer science, and its impact is growing even in scientific disciplines beyond computer science. Time complexity is the most important measure of computational complexity. It measures the number of computer instructions performed (the amount of computer work). Time complexity is usually represented by a function of the input size.

The main goal of complexity theory is classifying algorithmic problems with respect to the amount of work (to the computational complexity) that is sufficient and necessary for computing their correct solutions. The final global aim is to define the border between practical solvability and intractability, and so to classify algorithmically solvable problems into practically solvable and practically unsolvable. The first approximation of the limit of practical solvability proposes that we consider those problems practically solvable that can be solved in polynomial-time with respect to the input length. A problem for whose solution no polynomial-time algorithm exists is considered to be hard (practically unsolvable). Complexity theory teaches us that there exist problems of an arbitrarily large complexity. For instance, there exist computing problems that can be solved only by algorithms having time complexity at least 2^{2^n}.

To prove the nonexistence of efficient algorithms for solving concrete computing problems is the hardest research problem in the investigation of the laws of information processing. We are not able to solve this fundamental problem of computer science because we lack sufficiently powerful mathematical methods for this purpose. Therefore, one currently uses incomplete arguments for arguing about the hardness of some problems. One again uses the reduction method in order to define a large class of equally hard problems in the sense that each problem of this class can be efficiently reduced to any other problem of this class. The consequence is that either all problems of this class are efficiently solvable or none of them is efficiently solvable. The problems in this class are called NP-hard problems. For none of the NP-hard problems does one know an efficient polynomial algorithm, and all algorithms discovered for NP-hard problems are of exponential complexity. Because the existence of an efficient algorithm for any of the NP-hard problems would mean the efficient solvability of all NP-hard problems, and one does not believe in efficient solvability of all these problems, we consider NP-hardness as a reliable argument for the computational hardness of a problem.

The art of algorithm design lies in searching for solutions to hard problems. The most important discovery is that many hard problems are very unstable with respect to the amount of work necessary to solve them (to their hardness). A small (almost negligible[8]) change of the problem formulation or a small reduction of the constraints on feasible solutions may cause an unbelievably huge jump from a physically unexecutable amount of computer work to a matter of a few seconds on a common PC. To discover and use this sensitivity of hard problems is at the heart of current algorithmics [Hro04a], which has the goal of making the solution of hard problems affordable for industry and, in the end, for the development of an information society.

Solutions to Some Exercises

Exercise 5.2 Applying the distributive law three times, one can rewrite a polynomial of degree 4 as follows:

[8] from the practical point of view

$$\begin{aligned} & a_4 \cdot x^4 + a_3 \cdot x^3 + a_2 \cdot x^2 + a_1 \cdot x + a_0 \\ = & \left[a_4 \cdot x^3 + a_3 \cdot x^2 + a_2 \cdot x^1 + a_1\right] \cdot x + a_0 \\ = & \left[(a_4 \cdot x^2 + a_3 \cdot x + a_2) \cdot x + a_1\right] \cdot x + a_0 \\ = & \left[((a_4 \cdot x + a_3) \cdot x + a_2) \cdot x + a_1\right] \cdot x + a_0 \ . \end{aligned}$$

If one applies the resulting formula for computing the value of the polynomial for given values of a_4, a_3, a_2, a_1, a_0, and x, then only 4 multiplications and 4 additions are performed.

Exercise 5.4

(a) The value of x^6 can be computed by 3 multiplications using the following strategy:

$$I \leftarrow x \cdot x \ , \ J \leftarrow I \cdot I \ , \ Z \leftarrow J \cdot I.$$

We see the idea starting with computing $x^2 = x \cdot x$ using one multiplication and continuing by computing $x^6 = x^2 \cdot x^2 \cdot x^2$ using two multiplications.

(b) The value of x^{64} can be computed using 6 multiplications as follows:

$$x^{64} = (((((x^2)^2)^2)^2)^2)^2$$

This strategy can be implemented as follows:

$$\begin{aligned} & I \leftarrow x \cdot x \ , \ I \leftarrow I \cdot I \ , \ I \leftarrow I \cdot I \ , \\ & I \leftarrow I \cdot I \ , \ I \leftarrow I \cdot I \ , \ I \leftarrow I \cdot I \ . \end{aligned}$$

(c) One can propose the following strategy for computing the value of x^{18}.

$$x^{18} = (((x^2)^2)^2)^2 \cdot x^2 \ .$$

It can be implemented as follows:

$$\begin{aligned} & I \leftarrow x \cdot x \ , \ J \leftarrow I \cdot I \ , \ J \leftarrow J \cdot J \ , \\ & J \leftarrow J \cdot J \ , \ Z \leftarrow I \cdot J \ . \end{aligned}$$

(d) The value of x^{32} can be computed in the following way:

$$\begin{aligned} x^{45} &= x^{32} \cdot x^8 \cdot x^4 \cdot x \\ &= ((((x^2)^2)^2)^2)^2 \cdot ((x^2)^2)^2 \cdot (x^2)^2 \cdot x \end{aligned}$$

using the following implementation:

$$\begin{aligned} & I_2 \leftarrow x \cdot x \ , \quad I_4 \leftarrow I_2 \cdot I_2 \ , \quad I_8 \leftarrow I_4 \cdot I_4 \ , \\ & I_{16} \leftarrow I_8 \cdot I_8 \ , \ I_{32} \leftarrow I_{16} \cdot I_{16} \ , \ Z \leftarrow I_{32} \cdot I_8 \ , \\ & Z \leftarrow Z \cdot I_4 \ , \quad Z \leftarrow Z \cdot x \end{aligned}$$

We see that 8 multiplications are sufficient.

One can even compute the value x^{45} with 7 multiplications only by using the following strategy:
$x^2 = x \cdot x$, $x^3 = x^2 \cdot x$, $x^6 = x^3 \cdot x^3$, $x^9 = x^6 \cdot x^3$,
$x^{18} = x^9 \cdot x^9$, $x^{36} = x^{18} \cdot x^{18}$, $x^{45} = x^{36} \cdot x^9$

Exercise 5.9 If a mathematician has to solve the linear equation $a + bx = c + dx$, she or he can simplify the equality in the following way:

$$\begin{aligned} a + bx &= c + dx \quad | -c \\ a - c + bx &= dx \qquad | -dx \\ (a-c) + bx - dx &= 0 \\ (a-c) + (b-d)\cdot x &= 0 \qquad \{\text{following the distributive law}\} \end{aligned}$$

Now, the linear equation has the simplified form, for which we already have an algorithm. In Fig. 5.11 we present the graphic representation of the corresponding efficient reduction.

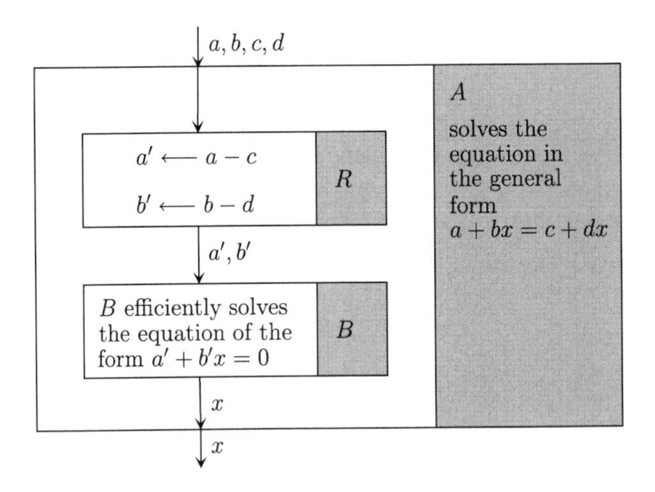

Fig. 5.11

The reduction R is efficient because two subtractions satisfy in order to transform a general linear equation into the form $a' + b'x = 0$. Since $a' + b'x = 0$ is only a conversion (another representation) of the equation $a + bx = c + dx$, both equations have the same solution. Therefore, the output of the program B is also the output of the algorithm A for solving general linear equations.

Exercise 5.10 If there is no supervisor at the crossroad K_4, there must be 4 supervisors at the other ends of the 4 streets leading from K_4 in order to guarantee control over these four streets. We see that these 4 supervisors supervise the complete road network (Fig. 5.5).

Exercise 5.17 For the 7 crossroads and at most 3 supervisors one gets first the following linear inequation:

$$x_1 + x_2 + x_3 + x_4 + x_5 + x_6 + x_7 \leq 3 \ .$$

For 8 streets one gets the following 8 inequalities:

$$x_1 + x_5 \geq 1 \ , \ x_1 + x_6 \geq 1 \ , \ x_2 + x_5 \geq 1 \ , \ x_2 + x_6 \geq 1 \ ,$$

$$x_3 + x_6 \geq 1 \ , \ x_3 + x_7 \geq 1 \ , \ x_4 + x_6 \geq 1 \ , \ x_4 + x_7 \geq 1 \ .$$

Because of the first inequality, we are allowed to assign the value 1 to at most 3 unknowns. The choices $x_5 = x_6 = x_7 = 1$ and $x_1 = x_2 = x_3 = x_4 = 0$ ensures that all 8 "street inequations" are satisfied.

Chance favours only those whose spirit has been prepared already, those unprepared cannot see the hand stretched out to them by fortune.

Louis Pasteur

Chapter 6

Randomness in Nature and as a Source of Efficiency in Algorithmics

6.1 Aims

In this chapter we aim to introduce one of the miracles of computer science. This marvel is based on a successful application of randomness in the design of efficient algorithms for apparently intractable problems.

To perform and to understand this magic one has to deal with the meaning of the notion of "randomness" first. Hence, we follow the history of science which is full of controversial discussions about the existence of true randomness in order to get a feeling for what can be viewed as true randomness and what is only apparent randomness.

J. Hromkovič, *Algorithmic Adventures*, DOI 10.1007/978-3-540-85986-4_6,

Taking that generated view about randomness we directly move to algorithmics in order to apply it for solving computationally hard problems. Here we see a miracle. Can you imagine, at first glance, that systems with a randomized control can achieve their goals billions of times faster than any deterministic system can? Do you believe that one has to pay for this huge speed-up by risking making an error with probability[1] smaller than $1/10^{19}$ only? Is this really a serious risk? If one had started such a randomized system every second since the Big Bang (i.e. between 10^{18} to 10^{19} times), then one may have expected none of these many runs to make an error. In another words the probability of making at least one error in these more than 10^{18} experimental runs of this randomized system is smaller than the probability of reaching the correct goal in all these attempts.

Here we call attention to the fact that in practice randomized algorithms (algorithms with a random control) with very small error probability can be even more reliable than their best deterministic counterparts. What do we mean by this? Theoretically all deterministic algorithms may not err. But the nature of the story is that deterministic programs are not absolutely reliable because during their runs on a computer a hardware error may occur and then they may produce wrong results. Clearly, the probability of the occurrence of a hardware error grows proportionally with the running time of the program. Therefore a fast randomized algorithm can be more reliable than a slow deterministic algorithm. For instance, if a randomized algorithm computes a result in 10 seconds with an error probability $1/10^{30}$, then it is more reliable than a deterministic algorithm that computes the result in one week. Hence, moving from common (deterministic) algorithms and systems to randomized ones is not necessarily related to decreasing reliability. And losing an apparently absolute reliability may not hurt too much. Randomization, here we mean building controlling systems by using a clever combination of deterministic strategies with random decisions, became a new paradigm of algorithmics. Efficient randomized algorithms making one mistake in billions of

[1] The risk of computing a wrong result is called the error probability.

applications on average are very practical and enable us to solve problems considered intractable for deterministic computations.

After presenting a magical example to convince ourselves about the power and the usefulness of randomness we do not leave the reader in her or his amazement. We will try to give the reader at least a feel for why such unbelievably elegant and wonderful ways for solving problems exist. We get a deeper view into the nature of the success of randomized systems and recognize some paradigms for the design of randomized algorithms. And, maybe, at the very end we can even encourage you to use the obtained knowledge in your everyday life and so to become a magician.

6.2 What Is Randomness and Does There Exist True Randomness?

In the first chapter we realized that the creation of notions is fundamental for the foundation and the development of scientific disciplines. The notion of **randomness** is one of the few most fundamental and most discussed notions of science. Philosophers, mathematicians, physicists, biologists and chemists discussed the existence of randomness and its possible role for thousands of years. Mathematicians created probability theory as an instrument for investigating random events. Computer scientists study why and when it is profitable to use randomness and design randomized algorithms. Engineers apply the obtained knowledge to build efficient advantageous randomized systems.

The notion of randomness is strongly related to another two fundamental notions: **determinism** and **nondeterminism**. The deterministic view of the world is based on the paradigm of causality. Every event has a cause, i.e., every event is a consequence of a cause. Every cause has its effect (result) and this result is a cause of further effects, etc. Following the principle of causality, if one knows the correct state of the Universe and all natural laws, one can completely predict the future. The whole development is viewed as a chain of causes and their effects. By contrast, nondeterminism means that a cause can have several different consequences (effects) and there is no law determining which one must occur. In another words you can learn all possible results of a cause but only one of these results may occur and there is no possibility to predict which one.

The notion of randomness is strongly related to nondeterminism. Following the definitions used in dictionaries

an event is considered to be ***random***

if

it happens unpredictably.

Similarly,

an object is called ***random***

if

it is created without any pattern and any plan.

Clearly, the existence of random events (i.e., of the occurrence of randomness) contradicts the causality principle. Hence, the following question

> "Does there exist a true (objective) randomness or does one use this term only to model and describe events with unknown lawfulness?"

is fundamental for science. Philosophers and scientists have disputed the answer to this question since ancient times. Democritos believed that

randomness is the unknown,
and that Nature is determined in its fundamentals.

Thus, Democritos asserted that order conquers the world and this order is governed by unambiguous laws. Following Democritos's opinion, one uses the notion of "randomness" only in the subjective sense in order to veil one's inability to truly understand the nature of events and things. Hence the existence of the notion of randomness is only a consequence of the incompleteness of our knowledge. To present his opinion transparently, Democritos liked to use the following example. Two men agreed to send their slaves to bring water at the same time in order to cause the slaves to meet. The slaves really met at the source and said, "Oh, this is randomness that we have met."

In contrast to Democritos, Epicurus claimed that

> *randomness is objective,*
> *it is the proper nature of events.*

Thus, Epicurus claimed that there exists a true randomness that is completely independent of our knowledge. Epicurus's opinion was that there exist processes whose development is ambiguous rather than unambiguous, and an unpredictable choice from the existing possibilities is what we call randomness.

One could simply say, Epicurus was right because there are games of chance, such as rolling dice or roulette, that can have different outcomes, and the results are determined by chance. Unfortunately, the story is not so simple, and discussing gambling one gets support for the opinion of Democritos rather than for Epicurus' view on the nature of events. Rolling dice is a very complex activity, but if one knows the direction, the speed and the surface on which a die is tossed, then it may be possible to compute and predict the result. Obviously, the movement of the hand controlled by the human brain is too complex to allow us to estimate the values of all important parameters. But we may not consider the process of rolling a die as an objectively random process only because it is too complex for us to predict the outcome. The same is true of roulette and other games of chance. Physics also often uses random models to describe and analyze physical processes that are not inherently or necessarily random (and are sometimes clearly deterministic), but which are too complex to have a realistic possibility of modelling them in a fully deterministic way. It is interesting to note that based on this observation even Albert Einstein accepted the notion of randomness only in relation to an incomplete knowledge, and strongly believed in the existence of clear, deterministic laws for all processes in Nature[2].

Before the 20th century, the world view of people was based on causality and determinism. The reasons for that were, first, religion, which did not accept the existence of randomness in a world created by God[3], and, later, the optimism created by the success of natural sciences and mechanical engineering in the 19th century, which gave people hope that everything could be discovered, and everything discovered could be explained by deterministic causalities of cause and resulting effect[4].

This belief in determinism also had emotional roots, because people connected randomness (and even identified it) with chaos, uncertainty, and unpredictability, which were always related to fear; and so the possibility of random events was rejected. To express the strongly negative connotation of randomness in the past, one can consider the following quotation of Marcus Aurelius:

> *There are only two possibilities,*
> *either a big chaos conquers the world,*
> *or order and law.*

Because randomness was undesirable, it may be not surprising that philosophers and researchers performed their investigations without allowing for the existence of random events in their concepts or even tried to prove the nonexistence of randomness by focusing on deterministic causalities. Randomness was in a similarly poor situation with Galileo Galilei, who claimed that the Earth is not a fixed center of the whole Universe. Though he was able to prove his claim using experimental observations, he had no chance to convince people about it because they were very afraid

[2] "God does not roll dice" is a famous quotation of Albert Einstein. The equally famous reply of Niels Bohr is, "The true God does not allow anybody to prescribe what He has to do."

[3] Today we know that this strong interpretation is wrong and that the existence of true randomness does not contradict the existence of God.

[4] Take away the cause, and the effect must cease.

of such a reality. Life in the medieval world was very hard, and so people clung desperately to the very few assurances they had. And the central position of the Earth in the Universe supported the belief that poor Man is at the center of God's attention. The terrible fear of losing this assurance was the main reason for the situation, with nobody willing to verify the observations of Galileo Galilei. And "poor" randomness had the same trouble gaining acceptance[5].

Finally, scientific discoveries in the 20th century (especially in physics and biology) returned the world to Epicurus' view on randomness. The mathematical models of evolutionary biology show that random mutations of DNA have to be considered a crucial instrument of evolution. But, the essential reason for accepting the existence of randomness was one of the deepest discoveries in the history of science: the theory of quantum mechanics. The mathematical model of the behavior of particles is related to ambiguity, which can be described in terms of random events. All important predictions of this theory were proved experimentally, and so some events in the world of particles are considered to be truly random events. To accept randomness (or random events) it is very important to overcome the restrictive interpretation of randomness, identifying it with chaos and uncertainty. A very elegant, modern view on randomness is given by the Hungarian mathematician Alfréd Rényi:

> *Randomness and order do not contradict each other;*
> *more or less both may be true at once.*
> *Randomness controls the world*
> *and due to this in the world there are order and law,*
> *which can be expressed in measures of random events*
> *that follow the laws of probability theory.*

For us, as computer scientists, it is important to realize that there is also another reason to deal with randomness beyond the "mere"

[5] One does not like to speak about emotions in the so-called exact (hard) sciences, but this is a denial of the fact that the emotions of researchers (the subjects in the research) are aggregates of development and progress.

modelling of natural processes. Surprisingly, this reason was formulated over 200 years ago by the great German poet Johann Wolfgang von Goethe as follows:

> *The tissue of the world*
> *is built from necessities and randomness;*
> *the intellect of men places itself between both*
> *and can control them;*
> *it considers the necessity*
> *as the reason for its existence;*
> *it knows how randomness can be*
> *managed, controlled, and used...*

In this context we may say that Johann Wolfgang von Goethe was the first "computer scientist" who recognized randomness as a useful source for performing certain activities. The use of randomness as a resource of an unbelievable, phenomenal efficiency is the topic of this chapter. We aim to convince the reader that it can be very profitable to design and implement randomized algorithms and systems instead of completely deterministic ones. This realization is nothing other than the acceptance of Nature as teacher. It seems to be the case that Nature always uses the most efficient and simplest way to achieve its goal, and that randomization of a part of the control is an essential element of Nature's strategy. Computer science practice confirms this point of view. In many everyday applications, simple randomized systems and algorithms do their work efficiently with a high degree of reliability, and we do not know any deterministic algorithms that would do the same with a comparable efficiency. We even know of examples where the design and use of deterministic counterparts of some randomized algorithms is beyond physical limits. This is also the reason why currently one does not relate tractability (practical solvability) with the efficiency of deterministic algorithms, but with efficient randomized algorithms.

To convince the reader of the enormous usefulness of randomization, the next section presents a randomized protocol that solves a concrete communication task within communication complexity that is substantially smaller than the complexity of the best possible deterministic protocol.

We close this section by calling attention to the fact that we did not give a final answer to the question of whether or not true randomness exists, and it is very improbable that science will be able to answer this question in the near future. The reason for this pessimism is that the question about the existence of randomness lies in the very fundamentals of science, i.e., at the level of axioms, and not at the level of results. And, at the level of axioms (basic assumptions), even the exact sciences like mathematics and physics do not have any generally valid assertions, but only assumptions expressed in the form of axioms. The only reason to believe in axioms is that they fully correspond to our experience and knowledge. As mentioned already in Chapter 1, an example of an axiom of mathematics (viewed as a formal language of science) is assuming that our way of thinking is correct, and so all our formal arguments are reliable. Starting with the axioms, one constructs the building of science very carefully, in such a way that all results achieved are true provided the axioms are valid. If an axiom is shown to be not generally valid, one has to revise the entire theory built upon it[6].

Here, we allow ourselves to believe in the existence of randomness, and not only because the experience and knowledge of physics and evolutionary theory support this belief. For us as computer scientists, the main reason to believe in randomness is that randomness can be a source of efficiency. Randomness enables us to reach aims incomparably faster, and it would be very surprising for us if Nature left this great possibility unnoticed.

[6] Disproving the general validity of an axiom should not be considered a "tragedy." Such events are part of the development of science and they are often responsible for the greatest discoveries. The results built upon the old, disproved axiom usually need not be rejected; it is sufficient to relativize their validity, because they are true in frameworks where the old axiom is valid.

6.3 The Abundance of Witnesses Is Support in Shortage or Why Randomness Can Be Useful

The aim of this section is to present a task that can be solved efficiently by applying randomness despite the known fact that this task cannot be solved efficiently in any deterministic way. Here we design a randomized system for solving the mentioned task in such a simple way that anybody can use it in order to get a first convincing experience of the extraordinary power of randomized control.

What is the exact meaning of a **randomized** system or randomized algorithm (program)? In fact one allows two different basic transparent descriptions. Each deterministic program executes on a given input an unambiguously given sequence of computing steps. We say that the program together with its input unambiguously determines a unique computation. A randomized program (system) can have many different computations on the same input, and which one is going to be executed is decided at random. In what follows we present two possibilities for the performance of the random choice from possible computations.

1. The program works in the deterministic way except for a few special places, in which it may flip a coin. Depending on the result of coin flipping (head or tail), the program takes one of the possible two ways to continue in the work on the input.

   To insert this possibility of random choice into our programming language TRANSPARENT, it is sufficient to add the following instruction:

   ```
   Flip a coin. If ''head'', goto i else goto j.
   ```

   In this way the program continues to work in the i-th row if "head" was the result of flipping coin and it continues in the j-th row if "tail" was the result of flipping coin.

2. The randomized algorithm has at the beginning a choice of several deterministic strategies. The program randomly chooses one of these strategies and applies it on the given input. The

rest of the computation is completely deterministic. The random choice is reduced to the first step of computation. For each new problem instance the algorithm chooses a new strategy at random.

Since the second way of modelling randomized systems is simpler and more transparent than the first one, we use it for the presentation of our exemplary application of the power of randomness.

Let us consider the following scenario.

We have two computers R_I and R_{II} (Fig. 6.1) that are very far apart. The task is to manage the same database on the computers. At the beginning both have a database with the same content. In the meantime the contents of these databases dynamically develop in such a way that one tries to perform all changes simultaneously on both databases with the aim of getting the same database, with complete information about the database subject (for instance, genome sequences), in both locations. After some time, one wants to check whether this process is successful, i.e., whether R_I and R_{II} contain the same data.

We idealize this verification task in the sense of a mathematical abstraction. We consider the contents of the databases of R_I and R_{II} as sequences of bits. Hence, we assume that the computer R_I has a sequence

$$x = x_1 x_2 x_3 \ldots x_{n-1} x_n$$

of n bits and the computer R_{II} has the n-bits sequence

$$y = y_1 y_2 y_3 \cdots y_{n-1} y_n .$$

The task is to use a communication channel (network) between R_I and R_{II} in order to verify whether $x = y$ (Fig. 6.1).

To solve this task one has to design a communication strategy, called a **communication protocol** in what follows. The **complexity of communication** and so the complexity of solving the verification problem is measured in the number of bits exchanged between R_I and R_{II} through the communication channel.

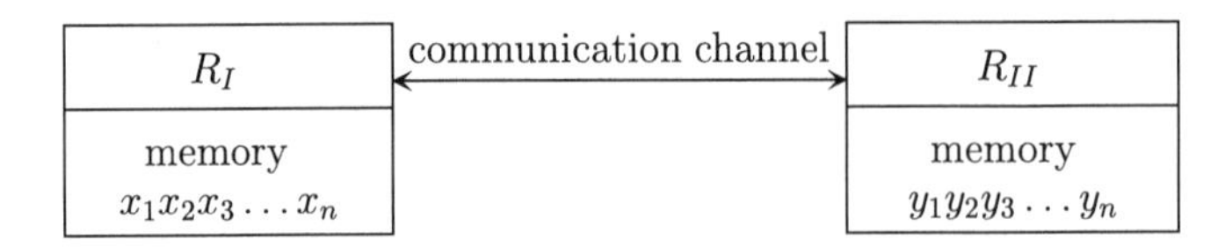

Fig. 6.1

Unfortunately, the provable fact is that any communication protocol for solving this task cannot do better for most possible inputs x and y than to exchange n bits. This means that the naive protocol based on sending all n bits (the whole x) of R_I to R_{II} and asking R_{II} to perform the comparison of x and y bit by bit is optimal. No multiple message exchange and no clever computation of R_I and R_{II} can help to decrease the communication complexity. If n is very large, for instance $n = 10^{16}$ (this corresponds to the memory size of 250000 DVDs), then the communication cost is too high. Moreover to reliably transfer 10^{16} bits without flipping or losing any particular bit using current Internet technologies is an almost unrealistic undertaking.

In what follows we design a randomized communication protocol that solves our "equality problem" within

$$4 \cdot \lceil \log_2(n) \rceil$$

communication bits[7]. We see that one can save an exponential complexity in this way. For instance, for $n = 10^{16}$ one needs to send only 256 bits instead of 10^{16} bits!

For the presentation of the randomized communication strategy it is more transparent to speak about a comparison of two numbers instead of the comparison of two bit sequences. Therefore we consider the sequences $x = x_1 \dots x_n$ and $y = y_1 \dots y_n$ of bits as binary representations of the integers.

[7] For a real number a, $\lceil a \rceil$ is the next larger integer. For instance, for $a = 4.15$, $\lceil a \rceil = 5$.

$$\mathbf{Number}(\boldsymbol{x}) = \sum_{i=1}^{n} 2^{n-i} \cdot x_i$$

and

$$\mathbf{Number}(\boldsymbol{y}) = \sum_{i=1}^{n} 2^{n-i} \cdot y_i$$

If these formulas do not mean anything to you anymore, and you have forgotten this secondary-school subject, you do not need to worry about this. The only important thing is to know that

Number(x) is a natural number represented by x, and that

$$0 \leq \text{Number}(x) \leq 2^n - 1.$$

Analogously,

$$0 \leq \text{Number}(y) \leq 2^n - 1.$$

For $n = 10^{16}$, these numbers can be really huge. Already for $n = 10^6 = 1000000$, the numbers can be around

$$2^{1000000}$$

and consist of about 300000 decimal digits.

Clearly, x and y are identical if and only if Number(x) = Number(y). Hence, we may focus on comparing Number(x) with Number(y) in our randomized protocol.

Exercise 6.1 For those who want to understand everything in detail. The sequence 10110 represents the integer

$$\begin{aligned}\text{Number}(10110) &= 1 \cdot 2^4 + 0 \cdot 2^3 + 1 \cdot 2^2 + 1 \cdot 2^1 + 0 \cdot 2^0 \\ &= 1 \cdot 16 + 0 \cdot 8 + 1 \cdot 4 + 1 \cdot 2 + 0 \cdot 1 \\ &= 16 + 4 + 2 = 22.\end{aligned}$$

What number is represented by the bit sequence 101001? What is the binary representation of the integer 133?

The randomized communication protocol for inputs $x_1 \ldots x_n$ and $y_1 \ldots y_n$ corresponds to a random choice of a collection of deterministic communication strategies. The size of this collection equals the number of primes smaller than n^2.

In what follows, for each positive integer $m \geq 2$, we denote by

$$\mathrm{PRIM}(m) = \{p \text{ is a prime} \mid p \leq m\}$$

the set of all primes in the interval from 1 to m and by

$$\mathrm{Prim}(m) = |\mathrm{PRIM}(m)|$$

the number of primes in PRIM(m).

In what follows we denote by

$$r = a \bmod b$$

the residue of the division $a : b$. For instance $2 = 14 \bmod 3$, because $14 : 3 = 4$ and the remainder is $r = 14 - 3 \cdot 4 = 2$.

If one considers the division $a : b$ in the framework of integers only, and the result is c and the remainder is $r < b$, then one can write

$$a = b \cdot c + r \ .$$

In our example for $a = 14$ and $b = 3$, the result of the division is $c = 4$ and the remainder is $r = 2$. Hence, one can write

$$14 = 3 \cdot 4 + 2,$$

where $r = 2 < 3 = b$.

Now, we are ready to present a randomized communication protocol for the comparison of x and y (better to say for comparing Number(x) with Number (y)).

Randomized communication protocol WITNESS for identity verification.

Initial situation: The computer R_I has n bits $x = x_1 x_2 \ldots x_n$ (i.e., an integer Number(x), $0 \leq$ Number$(x) \leq 2^n - 1$).
The computer R_{II} has n bits $y = y_1 y_2 \ldots y_n$ (i.e., an integer Number(y), $0 \leq$ Number$(y) \leq 2^n - 1$).

Phase 1: R_I chooses a prime p from PRIM(n^2) at random. Every prime from PRIM(n^2) has the same probability $1/\text{Prim}(n^2)$ to be chosen.

Phase 2: R_I computes the integer

$$s = \text{Number}(x) \bmod p$$

(i.e., the remainder of the division Number$(x) : p$) and sends the binary representations of

$$s \text{ and } p$$

to R_{II}.

Phase 3: After reading s and p, R_{II} computes the number

$$q = \text{Number}(y) \bmod p.$$

If $q \neq s$, then R_{II} outputs "unequal".
If $q = s$, then R_{II} outputs "equal".

Before analyzing the communication complexity and the reliability of WITNESS, let us illustrate the work of the protocol for a concrete input.

Example 6.1 Let $x = 01111$ and $y = 10110$.
Hence,

$$\text{Number}(x) = 1 \cdot 2^3 + 1 \cdot 2^2 + 1 \cdot 2^1 + 1 \cdot 2^0 = 8 + 4 + 2 + 1 = 15,$$
$$\text{Number}(y) = 1 \cdot 2^4 + 1 \cdot 2^2 + 1 \cdot 2^1 = 16 + 4 + 2 = 22$$

and

$n = 5.$

Consequently $n^2 = 25$ and

$$\text{PRIM}(n^2) = \{2, 3, 5, 7, 11, 13, 17, 19, 23\}.$$

In Phase 1 the communication protocol WITNESS has the random choice from the collection of 9 primes of PRIM(25) and in this way it chooses one of 9 possible deterministic protocols.

Assume that R_I chooses the prime 5. In Phase 2 computer R_{II} computes

$s = 15 \bmod 5 = 0$

and sends the integers $p = 5$ and $s = 0$ to R_{II}. Then R_{II} computes

$q = 22 \bmod 5 = 2.$

Since $2 = q \neq s = 0$, the computer R_{II} gives the correct answer

"x and y are unequal".

Assume now that R_I chooses the prime 7 from PRIM(25) at random. Then the computer R_I computes

$s = 15 \bmod 7 = 1$

and sends the integers $p = 7$ and $s = 1$ to R_{II}. Then R_{II} computes

$q = 22 \bmod 7 = 1.$

Since $q = s$, the computer R_{II} gives the wrong answer

"x and y are equal".

We immediately observe that WITNESS may err because some of the random choices of primes can lead to a wrong answer. □

Exercise 6.2 Does there exist a prime in PRIM(23) different from 7, whose choice results in a wrong answer of WITNESS for $x = 01111$ and $y = 10110$?

Exercise 6.3 Let $x = y = 100110$ be the inputs. Does there exist a choice of a prime from PRIM(6^2) such that WITNESS provides the wrong answer "$x \neq y$"?

Exercise 6.4 Consider $x = 10011011$ and $y = 01010101$ as inputs. Estimate for how many primes from PRIM(8^2) WITNESS computes the wrong answer "$x = y$". What is the number of primes from PRIM(64) for which WITNESS provides the right answer "$x \neq y$"?

First, we analyze the communication complexity of WITNESS. The natural numbers Number(x) and Number(y) have the length of their binary representations bounded by n and so they are from the interval from 0 to $2^n - 1$. To save communication bits the communication of WITNESS consists of sending two integers smaller than n^2, namely p and the remainder of the division Number(x) : p. A natural number smaller than n^2 can be binary represented by

$$\lceil \log_2 n^2 \rceil \leq 2 \cdot \lceil \log_2 n \rceil$$

bits. The symbols $\lceil$ and $\rceil$ are used to denote the rounding to the next larger integer. For instance, $\lceil 2.3 \rceil = 3$, $\lceil 7.001 \rceil = 8$ and $\lceil 9 \rceil = 9$.

Since in WITNESS two numbers p and s are communicated, R_I can use exactly[8] $2\lceil \log_2 n \rceil$ bits for the representation of each one. In this way the number of communication bits for inputs of length n is always exactly

$$4 \cdot \lceil \log_2 n \rceil \, .$$

Let us see what that means for $n = 10^{16}$. As already mentioned, the best deterministic protocol cannot avoid the necessity of communicating at least

$$10^{16} \text{ bits}$$

for some inputs. Our protocol WITNESS always works within

$$4 \cdot \lceil \log_2(10^{16}) \rceil \leq 4 \cdot 16 \cdot \lceil \log_2 10 \rceil = 256 \text{ communication bits.}$$

The gap between communicating 10^{16} bits and 256 bits is huge. Even if it is possible to communicate 10^{16} bits in a reliable way the costs of sending 256 bits and of sending 10^{16} bits are incomparable. For this unbelievably large saving of communication costs we pay

[8] If the binary representation of p or s is shorter than $2\lceil \log_2 n \rceil$ one can add a few 0's at the beginning of the representation in order to get exactly this length.

by losing the certainty of always getting the correct result. The question of our main interest now is:

How large is the degree of unreliability we have used to pay for the huge reduction of the communication complexity?

The degree of uncertainty of computing the correct result is called the **error probability** in algorithmics. More precisely, the **error probability for two input strings x and y** is the probability

$$\mathbf{Error}_{WITNESS}(x, y)$$

that WITNESS computes a wrong answer on inputs x and y. For different input pairs (x, y) (i.e., for different initial situations) the error probabilities may differ. Our aim is to show that the error probability is very small for all[9] possible inputs x and y.

"What is the error probability for a given x and y and how can one estimate it?"

The protocol WITNESS chooses a prime for PRIM(n^2) for input strings x and y of length n at random. These choices decide whether the protocol provides the correct answer or not. Therefore we partition the set PRIM(n^2) into two parts. All primes that provide the correct answer for the input instance (x, y) we call

good for (x, y) .

The prime 5 is good for $(01111, 10110)$ as we have seen in Example 6.1.

The other primes whose choices result in the wrong answer for the problem instance (x, y) we call

bad for (x, y) .

The prime 7 is bad for $(01111, 10110)$ as was shown in Example 6.1.

[9] In computer science we have very high requirements on the reliability of randomized algorithms and systems. We force a high probability of getting the correct result for every particular problem instance. This is in contrast to the so-called stochastic procedures, for which one only requires that they work correctly with high probability on a statistically significant subset of problem instances.

Since each of the $\text{Prim}(n^2)$ many $\text{PRIM}(n^2)$ has the same probability of being chosen, we have

$$\text{Error}_{WITNESS}(x, y) = \frac{\text{the number of bad primes for } (x, y)}{\text{Prim}(n^2)},$$

i.e., the error probability is the ratio between the number of bad primes for (x, y) in the urn (in $\text{PRIM}(n^2)$) and the number $\text{Prim}(n^2)$ of all primes in the urn. This is a simple consideration and we can convince the reader about it in a transparent way. Assume an urn with 15 balls. Assume that exactly 3 of these 15 balls are white. Then the probability of choosing a white ball at random is exactly $\frac{3}{15} = \frac{1}{5}$, if each ball has the same probability of being chosen. In another words, 20% of the balls in the urn are white and so one can expect that in a long sequence of repeated random experiments (choosing a ball at random and putting it back) a white ball is the result of the random choice in 20% of cases. Similarly, flipping a fair coin many times you may expect the number of flipped heads is approximately the same as the number of flipped tails. Analogously, if WITNESS chooses a prime from the 15 primes in $\text{PRIM}(7^2)$ for a given (x, y) at random and there are two bad primes for (x, y) among these 15 primes, the error probability for (x, y) is $2/15$.

The situation is depicted in Fig. 6.2. Our task is to show, for any instance (x, y) of our identity problem, that the set of bad primes for (x, y) is very small with respect to the size of $\text{PRIM}(n^2)$.

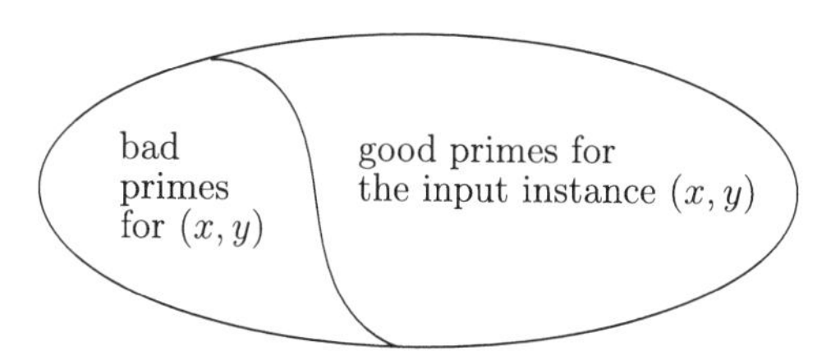

Fig. 6.2

How large is PRIM(m) for a positive integer m? One of the deepest and most important discoveries of mathematics[10] is the prime number theorem saying that

$$\text{Prim}(m) \text{ is approximately } \frac{m}{\ln m},$$

where $\ln m$ is the natural logarithm of m. The prime number theorem says that the primes are closely distributed among natural numbers, because one can find a prime approximately at each $(\ln m)$-th position. For our calculation we need only the known assertion

$$\text{Prim}(m) > \frac{m}{\ln m}$$

for all integers $m > 67$. Hence, we have

$$\text{Prim}(n^2) > \frac{n^2}{2 \ln n}$$

for all $n \geq 9$. Our next objective is to show that

> *for every problem instance (x,y), the number of bad primes for (x,y) is at most $n-1$,*

which is essentially smaller than $n^2/2\ln n$.

Investigating the error probability of WITNESS, we distinguish two cases with respect to the relation between x and y.

Case 1 $x = y$ and so the right answer is "equal".

If $x = y$, then Number(x) =Number(y). For all primes p the equality

$$s = \text{Number}(x) \bmod p = \text{Number}(y) \bmod p = q$$

holds. Hence, it does not matter which p from PRIM(n^2) is chosen, $s = q$ must hold. In other words, if one divides two equal numbers by the same prime p, the residues of these two divisions must be the same. Hence, the protocol WITNESS computes the right answer "equal" for all primes in PRIM(n^2). Therefore

$$\text{Error}_{WITNESS}(x, y) = 0$$

---

[10] more precisely, of number theory

for all strings $x = y$ (for all problem instances (x, y)).

From the analysis of Case 1 we learned that a wrong answer can occur only for problem instances (x, y) with $x \neq y$.

Case 2 $x \neq y$ and so the right answer is "unequal". As we have already fixed for $(01111, 10110)$ in Example 6.1, the error probability can be different from 0 for problem instances (x, y) with $x \neq y$. The random choice of $p = 7$ results in the wrong answer "equal" for $x = 01111$ and $y = 10110$.

In what follows we investigate the properties of bad primes for a given (x, y) in order to prove a general upper bound $n - 1$ on the number of bad primes for every problem instance (x, y) with $x \neq y$ and $|x| = |y| = n$.

A prime p is bad for (x, y) if the residues of the divisions $\text{Number}(x) : p$ and $\text{Number}(y) : p$ equal each other, i.e., if

$$s = \text{Number}(x) \bmod p = \text{Number}(y) \bmod p.$$

The equality

$$s = \text{Number}(x) \bmod p$$

means nothing other than

$$\text{Number}(x) = h_x \cdot p + s,$$

where h_x is the result of the division $\text{Number}(x) : p$ and $s < p$ is the remainder.

Analogously one can also write

$$\text{Number}(y) = h_y \cdot p + s,$$

where p is involved h_y times in $\text{Number}(y)$ and $s < p$ is the remainder[11].

Assume $\text{Number}(x) \geq \text{Number}(y)$ (in the opposite case when $\text{Number}(y) > \text{Number}(x)$ the analysis can be finished in an analogous way). We compute the integer $\text{Number}(x) - \text{Number}(y)$ as follows:

[11] For instance, for $x = 10110$ we have $\text{Number}(x) = 22$. For $p = 5$, $\text{Number}(x) = 22 = 4 \cdot 5 + 2$, where $h_x = 4$ and $s = 2$ is the remainder.

$$\begin{array}{r} \text{Number}(x) \\ -\text{Number}(y) \\ \hline \text{Dif}(x,y) \end{array} \qquad \begin{array}{r} h_x \cdot p + s \\ -h_y \cdot p - s \\ \hline h_x \cdot p - h_y \cdot p \end{array}$$

In this way we obtain:

$$\text{Dif}(x,y) = \text{Number}(x) - \text{Number}(y) = h_x \cdot p - h_y \cdot p = (h_x - h_y) \cdot p.$$

The final conclusion is that the prime p divides the number

$$\text{Dif}(x,y) = \text{Number}(x) - \text{Number}(y).$$

What did we learn from this calculation?

A prime p is bad for (x,y) if and only if p divides Dif(x,y).

Why is this knowledge helpful? First of all, we have found a fast way of recognizing bad primes.

Example 6.2 Assume R_I has the string $x = 1001001$ and R_{II} has the string $y = 0101011$, both strings of length $n = 7$. The task is to estimate the bad primes for $(x,y) = (1001001, 0101011)$.

First we estimate

$$\begin{aligned} \text{PRIM}(n^2) &= \text{PRIM}(49) \\ &= \{2,3,5,7,11,13,17,19,23,29,31,37,41,43,47\} \end{aligned}$$

as the set of primes from which WITNESS chooses a prime at random. Following our new observation, the bad primes for $(1001001, 0101011)$ are exactly those primes that divide the difference

$$\begin{aligned} \text{Dif}(1001001, 0101011) &= \text{Number}(1001001) - \text{Number}(0101011) \\ &= 73 - 43 = 30. \end{aligned}$$

We immediately see that 2, 3, and 5 are the only primes in PRIM(49) that divide 30, and so 2, 3, and 5 are the only bad primes for $(1001001, 0101011)$. In this way we obtain

$$\text{Error}_{WITNESS}(1001001, 0101011) = \frac{3}{\text{Prim}(49)} = \frac{3}{15} = \frac{1}{5}.$$

□

Exercise 6.5 Find all bad primes for the following problem instances.

(i) (01010, 11101)
(ii) (110110, 010101)
(iii) (11010010, 01101001).

Now, the remaining question is the following one:

> "How to use the knowledge that bad primes for (x, y) divides $\text{Dif}(x, y)$ to bound their number?"

Since both numbers $\text{Number}(x)$ and $\text{Number}(y)$ are smaller than 2^n, we have

$$\text{Dif}(x, y) = \text{Number}(x) - \text{Number}(y) < 2^n.$$

We show that a number smaller than 2^n cannot be divisible by more than $n - 1$ primes[12]. To do so we apply another famous theorem of mathematics, the so-called fundamental theorem of arithmetics. This theorem says that

> *each positive integer larger than 1 can be unambiguously expressed as a product of primes.*

For instance,

$$5940 = 2 \cdot 2 \cdot 3 \cdot 3 \cdot 3 \cdot 5 \cdot 11 = 2^2 \cdot 3^3 \cdot 5 \cdot 11$$

and so $2, 3, 5$, and 11 are the so-called prime factors of 5940. In other words, the number 5940 is divisible by exactly 4 primes $2, 3, 5$, and 11.

Let $p_1, p_2, p_3, \ldots, p_k$ be all prime factors of our number $\text{Dif}(x, y)$ and let $p_1 < p_2 < p_3 < \ldots < p_k$. In general, we see that $p_i > i$ for each i. Hence, we have

[12] Mathematicians would say a number smaller than 2^n can have at most $n-1$ prime factors.

$$\text{Dif}(x, y) = p_1^{i_1} \cdot p_2^{i_2} \cdot p_3^{i_3} \cdot \ldots \cdot p_k^{i_k}$$

for $i_j \geq 1$ for all j from 1 to k, and we can write

$$\begin{aligned} \text{Dif}(x, y) &\geq p_1 \cdot p_2 \cdot p_3 \cdot \ldots \cdot p_k \\ &> 1 \cdot 2 \cdot 3 \cdot \ldots \cdot k \\ &= k! \end{aligned}$$

Since $\text{Dif}(x, y) < 2^n$ and $\text{Dif}(x, y) > k!$, we obtain

$$2^n > k!\ .$$

Since $n! > 2^n$ for $n \geq 4$, k must be smaller than n and in this way we have obtained the stated aim that

$$k \leq n - 1,$$

i.e., that the number of prime factors of $\text{Dif}(x, y)$ is at most $n - 1$ and so the number of bad primes for (x, y) is at most $n - 1$.

Using this result we can upperbound the error probability of WITNESS on (x, y) for $x \neq y$ as follows. For every problem instance (x, y) of length $n \geq 9$,

$$\begin{aligned} \text{Error}_{WITNESS}(x, y) &= \frac{\text{the number of bad primes for } (x, y)}{\text{Prim}(n^2)} \\ &\leq \frac{n - 1}{n^2 / \ln n^2} \\ &\leq \frac{2 \ln n}{n}. \end{aligned}$$

Hence, the error probability of WITNESS for problem instances (x, y) with $x \neq y$ is at most $2 \ln / n$, which is for $n = 10^{16}$ at most

$$\frac{0.36841}{10^{14}}\ .$$

In Example 6.2 we saw that the error probability can be relatively high, 1/5. This is because the analyzed *error probability decreases with growing n*. Therefore, for small problem instances of a few thousand bits, one recommends to compare x and y in a deterministic way, since the costs for such a comparison are anyway small. Only for larger problem instances is it profitable to apply the randomized protocol WITNESS.

Exercise 6.6 What are the upper bounds on the error probability of WITNESS for input sizes $n = 10^3$ and $n = 10^4$? Up to which problem instance size would you recommend using the protocol WITNESS?

Exercise 6.7 Estimate the exact error probabilities of WITNESS for the following problem instances (x, y):

(i) $(00011011, 10101101)$
(ii) $(011000111, 000111111)$
(iii) $(0011101101, 0011110101)$.

Recently, we have seen a real miracle. A randomized system can efficiently and reliably perform tasks that are not tractable for deterministic systems. We know many hard problems that cannot be practically solved because the fastest known algorithms for them require an exponential complexity. Several such problems have to be solved daily and we are able to perform them only due to randomized algorithms. The whole area of e-commerce includes implementations of randomized algorithms in the software used. In fact, bad news for people searching for absolute reliability! But, for those who know the risk and can calculate it and reduce it, randomized systems provide what they are looking for.

Now it is time to at least partially reveal the nature behind the magic of randomization. We want to understand why randomization can cause such magnificent effects, and thus examine applications of randomization in algorithmics as a natural phenomenon.

We start our considerations by expressing doubts about the success of randomization presented above. We saw that the randomized protocol WITNESS works efficiently and with a high reliability. But is it really true that the best deterministic protocol for the identity problem cannot avoid sending almost n bits for most problem instances, and that for some problem instances the exchange of n bits is necessary? We did not prove it here. Why may one have doubts with respect to these claims? Intuitively, these assertions look suspicious.

Let us go into detail and look at Fig. 6.2. We rename the good primes for (x, y) to **witnesses of the difference between x and y** (for short).We say that a prime p is a **witness of "$x \neq y$"** or a **witness for (x, y)** if

$$\text{Number}(x) \bmod p \neq \text{Number}(y) \bmod p.$$

A prime q is a non-witness of "$x \neq y$" if

$$\text{Number}(x) \bmod p = \text{Number}(y) \bmod p.$$

Thus, the good primes for (x, y) are witnesses for (x, y) and the bad primes are non-witnesses for (x, y). Using these new terms one can view the work of the protocol WITNESS as searching for a witness of "$x \neq y$". If WITNESS chooses a witness of "$x \neq y$" at random, then WITNESS proves the fact "$x \neq y$" and reliably provides the correct answer "unequal". If the protocol chooses a non-witness for (x, y) with $x \neq y$ at random, one cannot avoid the wrong answer "equal". The protocol WITNESS works correctly with high probability, because almost all **candidates for witnessing** in PRIM(n^2) are witnesses. The group of non-witnesses is so small relative to Prim(n^2) that the probability of choosing a non-witness at random is very small. And now one can present the following doubt.

> *If almost all elements of the set of the candidates for witnessing are witnesses for (x, y), why is it not possible to find a witness among them in an efficient, deterministic way and then to solve the task within a short communication? To fix a witness for (x, y) in a deterministic way means to exchange the randomness for a deterministic control and so to design an efficient deterministic communication protocol for the comparison of x and y.*

How to explain that this at first glance plausible idea does not work? Clearly, we can present a formal mathematical proof that there is no efficient deterministic communication protocol for this task. But that is connected with two problems. First, despite the fact that this proof is not too hard for mathematicians and computer scientists, it requires knowledge that cannot be presupposed by a general audience. Secondly, understanding of the arguments of the proof does not necessarily help us to realize why the proposed idea does not work. You would only know that designing an efficient deterministic protocol for this task is impossible, but the main reason for that would remain hidden. Because of that we prefer to present only an idea in a way that increases our understanding of randomized control.

There is no efficient[13] deterministic algorithm for finding a witness for a given (x, y), because from the point of view of the computers R_I and R_{II}

> *the witnesses are "randomly" distributed among the candidates for witnessing.*

What does this mean? If you know x but only very little about y, you cannot compute a witness for (x, y), even when several attempts are allowed. This is the consequence of the fact that the witnesses are completely differently distributed in the candidate set $\mathrm{PRIM}(n^2)$ for distinct inputs of length n. There is no rule for computing a witness for a partially unknown input. Hence, from the point of view of R_I and R_{II} the soup in the pot containing all candidates (Fig. 6.3) looks as a chaos (a really chaotic mix of witnesses and non-witnesses) And this is the kernel of the success of randomness.

For several tasks one can solve the problem efficiently if a witness is available. In a real application, nobody provides a witness for us for free. But if one can build a set of candidates for witnessing in such a way that this set contains many witnesses, it is natural to search for witnesses at random. The chaotic distribution of witnesses is no obstacle for a random choice. But if the distribution of witnesses among candidates is really random, there does not exist any efficient deterministic procedure for finding a witness in this set.

[13] with respect to communication complexity

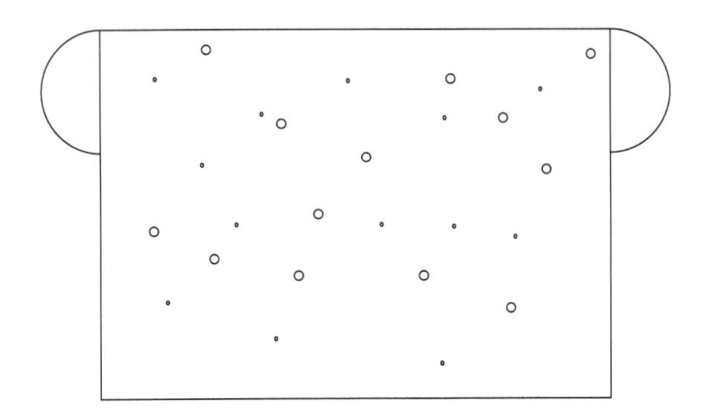

Fig. 6.3: A chaotic mix of witnesses and non-witnesses

Searching for efficient randomized algorithms one often applies the idea presented above. We try to find a kind of witness that fulfills the following requirements:

(i) If one has a witness for an input instance, then one can compute the correct result efficiently in a deterministic way.

(ii) If one has a candidate for witnessing, one can, for a given input, efficiently verify whether it is a witness or not.

(iii) There is an abundance of witnesses in the set of candidates.

Based on requirement (iii) this method for the design of randomized algorithms is called the method of the **abundance of witnesses**. Hence, to have many witnesses may be very useful. How one can even increase the abundance of witnesses is the topic of the next section.

6.4 What to Do When a Customer Requires a Very High Reliability?

The whole history of Man is strongly related to the search for certainty. Who are we? What are we looking for? What to do in order to get a guarantee for a "nice" or at least any future? Life and science educate us that the search for an absolute certainty is a nonsense that can even be dangerous. Striving for a non-existing absolute certainty means to do without many things and activities and often means to run up a blind alley. Frustration and depression are typical consequences of such meaningless effort. But if one accepts uncertainty and with it also randomness as an inevitable part of life and learn to live with this, instead of frustration one discovers new possibilities as a reward for giving up non-existing ideals. The blind alleys are not blind anymore and the future is open for a variety of possibilities. That was exactly the case in the example presented in the previous section. The situation looked hopeless, because each protocol solving the task has communication costs that are too large. Exchanging the hypothetical absolute reliability of deterministic protocols for a practical reliability of randomized protocols, we found a good solution for our communication task. In several similar situations one is unable to find witnesses that have as high abundance as in our example. Sometimes only a fraction of the candidates for witnessing are witnesses. In such cases the error probability can grow to an extent that is not acceptable in the applications. The aim of this section is to show how one can master such situations.

Let us start with our example of the randomized communication protocol for the equality of two strings. For $n = 10^{16}$, the error probability is at most $0.37 \cdot 10^{-14}$. Assume we have a customer for whom the reliability of the protocol is extremely important and so he is asking for a still higher degree of guarantee for the correct answer. For instance, the customer is saying:

> *"I require the reduction of the error probability to such a small value, that if one applies the protocol t times for t equal to the product of the age of the Universe in seconds and the number of protons in the Universe, then the probability of getting at least one wrong answer in one of these t applications is smaller than one in a million."*

For sure, that is a very strange requirement. Can we satisfy this exaggerated request of the customer without any huge increase in communication complexity? The answer is "YES", and in order to show how to do it we present a simple consideration from probability theory.

Assume we have a fair die. Rolling a die is considered a random experiment with 6 possible outcomes $1, 2, 3, 4, 5$, and 6, in which the probability of each concrete outcome is the same, i.e. $1/6$.

Assume that the only "bad" result for us is "1" and all other outcomes of the experiment are welcome. Hence, the probability of getting an undesirable outcome is $1/6$. Next, we change the experiment and the definition of an undesirable outcome. We consider rolling the die five times and the only undesirable outcome of this new experiment is to get "1" in all five rolls. In other words, we are satisfied if in at least one of these five simple experiments of rolling a die the outcome is different from "1". What is the probability of getting the undesirable outcome? To get the answer, we have to estimate the probability of rolling "1" five times, one behind another. Because all five die rolls are considered as independent[14] experiments, one calculates as follows. One "1" is rolled with probability $1/6$. Two "1" are rolled one behind another with probability

$$\frac{1}{6} \cdot \frac{1}{6} = \frac{1}{36}.$$

The probability of rolling five "1" one behind each other is exactly

$$\frac{1}{6} \cdot \frac{1}{6} \cdot \frac{1}{6} \cdot \frac{1}{6} \cdot \frac{1}{6} = \frac{1}{6^5} = \frac{1}{7776}.$$

What does this consideration have in common with the error probability of randomized communication protocols? A lot. One application of the randomized protocol can be viewed as a random experiment, in which the undesirable outcome corresponds to the random choice of a non-witness for a given input. Our analysis of the error probability of the randomized protocol shows the probability of getting this undesirable outcome, and so the wrong answer is at most

$$\frac{2 \ln n}{n}.$$

Now, we adapt the idea with the repeated die rolls. If one chooses 10 primes at random, then the only undesirable outcome is that all

[14] The exact meaning of the term independent is presented in [Hro04b].

10 primes are non-witnesses for the given input (x, y). If at least one prime is a witness for (x, y), one gets the correct answer with certainty. This consideration results in the following randomized communication protocol.

WITNESS (10)

Initial situation: Computer R_I has n bits $x = x_1 \ldots x_n$ and computer R_{II} has n bits $y = y_1 \ldots y_n$.

Phase 1: R_I chooses 10 primes

$$p_1, p_2 \ldots p_{10} \text{ from PRIM}(n^2) \text{ at random.}$$

Phase 2: For each i from 1 to 10 R_I computes

$$s_i = \text{Number}(x) \bmod p_i$$

and sends the binary representations of

$$p_1, p_2 \ldots p_{10}, s_1, s_2 \ldots s_{10}$$

to R_{II}.

Phase 3: After reaching the 20 numbers

$$p_1, p_2 \ldots p_{10}, s_1, s_2 \ldots s_{10},$$

the computer R_{II} computes

$$q_i = \text{Number}(y) \bmod p_i$$

for all $i = 1, 2 \ldots 10$.

If there exists at least one i from $1, 2, \ldots, 10$ such that $q_i \neq s_i$, then R_{II} knows with certainty that $x \neq y$ and outputs "unequal".

If $q_i = s_j$ for all 10 j from $1, 2, \ldots, 10$, then either $x = y$ or $x \neq y$ and none of the 10 primes chosen is a witness of "$x \neq y$". In this case the protocol outputs "$x = y$".

What do we observe? If $x = y$ then WITNESS(10) outputs the correct result "equal" with certainty, exactly as WITNESS did too. If $x \neq y$, then WITNESS(10) can provide the wrong output only if none of the 10 primes chosen is a witness for (x, y). It suffices that at least one of these 10 primes is a witness for (x, y), say p_4, and R_{II} knows that $x \neq y$ because $s_4 \neq p_4$ and so p_4 witnesses the difference between x and y. If the probability of choosing a non-witness for (x, y) in one attempt is at most $2 \ln /n$, then the probability of choosing 10 non-witnesses at random one behind the other is at most

$$\left(\frac{2 \ln n}{n}\right)^{10} = \frac{2^{10} \cdot (\ln n)^{10}}{n^{10}}.$$

For $n = 10^{16}$ this probability is at most

$$\frac{0.4714}{10^{141}}.$$

The product of the age of the Universe in seconds and the number of protons in the Universe is smaller than

$$10^{99}.$$

We omit presenting a mathematical calculation that shows that getting a wrong output in 10^{99} applications of WITNESS(10) is smaller than 1 in 10^{30}.

In this way we have reduced the error probability strongly below any reasonable limits and so we fulfill all reliability requirements that have ever been formulated in practice. And we paid very little for this wonderful gain. Computing with 10 primes instead of with one increases the communication costs by the multiplicative factor 10. Hence, the communication complexity of WITNESS(10) is

$$40 \cdot \lceil \log_2 n \rceil .$$

These communication costs are negligible. For instance, for $n = 10^{16}$, WITNESS(10) communicates

$$2560 \text{ bits}$$

only.

Since WITNESS(10) can be viewed as 10 repetitions of WITNESS, we say in the formal terms of algorithmics that

> *the complexity grows linearly with the number of repetitions (attempts to find a witness), while the error probability is reduced with an exponential speed-up in the number of repetitions.*

In fact, this situation belongs among the most favorable situations that one can have when searching for a solution to an algorithmic task.

Our randomized protocol WITNESS(10) is in fact more reliable than anything we could associate with the notion of reliability. There are two main reasons for that. First, already the protocol WITNESS provides a high degree of reliability. Secondly, the method of "repeated experiment" (of repeated random search for a witness) can essentially reduce the error probability even in cases when the probability of choosing a witness at random is small. The following exercises provide the reader with the possibility of applying the ideas presented above and so deepen one's imagination of the power of randomized computation.

Exercise 6.8 Let A be an algorithmic task for which every known algorithm has complexity at least 2^n. Let A be a decision problem for which only answers YES and NO are possible and so the task is to decide whether a problem instance x has a special property or not. Assume one has a kind of witness for x of size n with the following properties:

(i) If z is a witness for x, then one can verify in $10 \cdot n^2$ operations that x has the desired property. If z is not a witness for x, then the algorithm is unable to recognize whether x has the desired property or not. The fact that z is not a witness for x is recognized by the algorithm in time $10 \cdot n^2$, too.
(ii) At least half of the witness candidates are witnesses, i.e. the probability of choosing a witness at random is at least $1/2$.

The tasks are as follows:

1. Assume a randomized algorithm that performs 10 choices from the set of witness candidates at random in order to get a witness for the given input. Bound the error probability of this algorithm and analyze its computational complexity.

2. A customer is asking for an algorithm that solves the decision problem with error probability at most 1 in 10^9. How many repetitions of the original algorithm are necessary to reduce the error probability below this limit? What is the complexity of the resulting algorithm?

Exercise 6.9 Solve Exercise 6.8 if the assumption (i) is satisfied, and instead of (ii) we have the following assumptions:

(i) The portion of witnesses in the set of candidates is exactly 1/6.
(ii) The portion of witnesses in the set of candidates is only 1 in n, where n is the size of the problem instance (input).

6.5 What Are Our Main Discoveries Here?

If, for some tasks, one strongly requires to compute the correct results with absolute certainty, one can walk up a dead-end street. All algorithms that assume a theoretically absolute reliability demand so much computer work that they are intractable (not practicable). We know several computing tasks for which the best known algorithms need more time than the age of the Universe and more energy than the energy of the whole known Universe. To solve such problems is beyond the limits of physically doable. And then we discover that there is a magic way to solve these hard tasks. We do without the absolute reliability that does not exist anyway in practice and force a high probability of computing the correct result for each problem instance instead. We are allowed to require such a small error probability that one can speak about exchanging a hypothetical absolute reliability for practical certainty of getting the right result. Due to this fictitious reduction of our reliability requirements, we can save a huge amount of computer work. There exist problems for which a clever use of randomness causes a big jump from a physically undoable amount of work to a few seconds work on a common PC.

We learned two important paradigms for the design of randomized systems. The method of abundance of witnesses touches upon the deepest roots of the computational power of randomness. A witness for a problem instance is a piece of additional information that helps us to efficiently solve the problem instance that cannot be solved efficiently without it. The method promises successful applications if one can find a kind of witness, such that there is an abundance of witnesses in the set of candidates for witnessing. The idea is then to use repeated random sampling (choice) from the set of candidates in order to get a witness efficiently with high probability, although one does not know any efficient way of computing a witness using deterministic algorithms. The question of whether there are problems that can be solved efficiently by applying randomization, but which cannot be solved efficiently in a deterministic way, can be expressed as follows:

> *If all kinds of sets of candidates for witnessing for a hard problem have the witnesses randomly distributed in the candidate set then the problem is efficiently solvable using a randomized algorithm, but not efficiently solvable using deterministic algorithms.*

The second approach we applied for the design of randomized algorithms is the *paradigm of increasing success probability (of a fast reduction of error probability) by repeating several random computations on the same problem instance.* This paradigm is often called the *paradigm of amplification* (of the success probability). Repeating WITNESS 10 times results in the protocol WITNESS(10), whose error probability tends to 0 with an exponential speed in the number of repetitions, but whose complexity grows only linearly in the number of repetitions. The situation is not always so favorable, but usually we are able to satisfy the reliability requirements of customers without paying too much with additional computer work.

The textbook "The Design and Analysis of Randomized Algorithms" [Hro04b, Hro05] provides a transparent introduction to methods for designing efficient randomized systems. On one side it builds the intuition for the subject in small steps, and on the other side it consistently uses the rigorous language of mathematics in order to provide complete argumentation for all claims related to the quality of designed algorithms in terms of efficiency and error probability. The most exhaustive presentation of this topic is available in Motwani and Raghavan [MR95]. But this book is written for specialists and so it is not easily read by beginners. A delicacy for gourmets is the story about designing algorithms for primality testing by Dietzfelbinger [Die04], where the method of abundance of witnesses is applied several times. Further applications for designing randomized communication protocols are available in [KN97, Hro97]. Here, the mathematically provable advantages of randomness over determinism are presented. Unfortunately, because many highly nontrivial mathematical arguments are considered, these books are not accessible to a broad audience.

Solutions to Some Exercises

Exercise 6.2 The prime 7 is the only prime in PRIM(25) whose choice in the first step of WITNESS results in a wrong output "equal" for the input $x = 01111$ and $y = 10010$.

Exercise 6.5 (i) Let $x = 01010$ and $y = 11101$. Hence, $n = 5$ and we consider primes from $\text{PRIM}(5^2)$. The integers represented by x and y are:

$$\begin{aligned}\text{Number}(01010) &= 2^3 + 2^1 = 8 + 2 = 10 \\ \text{Number}(11101) &= 2^4 + 2^3 + 2^2 + 2^0 = 16 + 8 + 4 + 1 = 29.\end{aligned}$$

To find the bad primes from PRIM(25) for x and y, we do not need to test all primes from PRIM(25). We know that every bad prime for x and y divides the difference

$$\text{Number}(11101) - \text{Number}(01010) = 29 - 10 = 19$$

The number 19 is a prime and so 19 is the only prime in PRIM(25) that 19 divides. Hence, 19 is the only bad prime for 01010 and 11101.

Exercise 6.7 (i) To calculate the error probability of WITNESS for $x = 00011011$ and $y = 10101101$, we need to estimate the cardinality of $\text{PRIM}(8^2)$.

$$\text{PRIM}(64) = \{2, 3, 5, 7, 11, 13, 17, 19, 23, 29, 31, 37, 41, 43, 47, 53, 59, 61\}$$

and so $|\text{PRIM}(64)| = 18$. The integers represented by x and y are:

$$\begin{aligned}\text{Number}(x) &= 2^4 + 2^3 + 2^1 + 2^0 = 16 + 8 + 2 + 1 = 27 \\ \text{Number}(y) &= 2^7 + 2^5 + 2^3 + 2^2 + 2^0 = 128 + 32 + 8 + 4 + 1 = 173.\end{aligned}$$

Hence, $\text{Number}(y) - \text{Number}(x) = 173 - 27 = 146$. The factorization of the number 146 is:

$$146 = 2 \cdot 73\ .$$

Thus, the integers 2 and 73 are the only primes that divide the difference

$$\text{Number}(y) - \text{Number}(x) .$$

The prime 73 does not belong to PRIM(64) and so the prime 2 is the only bad prime for (x, y) in PRIM(64). Hence, the error probability of the protocol WITNESS for $(01010, 11101)$ is exactly $1/18$.

Exercise 6.8 The algorithm halts when it finds a witness for x. In the worst case, the algorithm performs 10 attempts and each attempt costs at most $10 \cdot n^2$ operations for checking whether the chosen candidate is a witness for x. Therefore the worst-case complexity of the randomized algorithm is $20 \cdot 10 \cdot n^2$. If the input x does not have the property we are searching for, the algorithm provides the right answer "NO" with certainty. If x has the property, then the algorithm can output the wrong answer "NO" only when it chooses 20 times a non-witness for x. We assume (assumption (ii) of the exercise) that the probability of choosing a non-witness at random in an attempt is at most $1/2$. Hence, the probability of choosing no witness in 20 attempts is at most

$$\left(\frac{1}{2}\right)^{20} = \frac{1}{2^{20}} = \frac{1}{1048576} \leq 0.000001 .$$

In this way we determined that the error probability is smaller than 1 in one million.

How many attempts are sufficient in order to reduce the error probability below $1/10^9$? We know that $2^{10} = 1024 > 1000$. Hence,

$$2^{30} = (2^{10})^3 > 1000^3 = 10^9 .$$

Thus, 30 attempts suffices for getting an error probability below 1 in one million.

From a genius idea all words can be removed.

Stanisław Jerzy Lec

Chapter 7

Cryptography, or How to Transform Drawbacks into Advantages

7.1 A Magical Science of the Present Time

In the 20th century physics was probably the most fascinating science. It brought discoveries that have essentially changed our view on the world. Many interpretations and applications of the theory of relativity or of quantum mechanics look like miracles. Things and events considered to be impossible became fundamental for building physical theories. We consider cryptography to be a magical science of the present time that transfers knowledge into magic. Do you believe that

- one can convince everybody about having a secret without revealing any small fraction of the secret?

J. Hromkovič, *Algorithmic Adventures*, DOI 10.1007/978-3-540-85986-4_7,

- one can send a message in cipher to several receivers in such a way that they can read it only if all receivers cooperate (i.e., they put their secrets together)?
- one can sign a document electronically in such a way that everybody is convinced about the authenticity of the signature but nobody can fake (emulate) it?
- one can agree on a common secret with another person in a public talk in such a way that no listener gets any information about the secret?

That and still much more is possible. Pure magic.

Communication is the generic term of the current era. It does not matter whether one uses mobile phones, SMS, e-mail or classical fax, phones or letters. Communication is present at every time and everywhere. A large amount of data is permanently flowing via the Internet. With the amount of data submitted, the difficulty of ensuring the privacy of all parties involved in communication increases. This is not only the problem of ensuring that only the proper receiver is able to read a message submitted to him. In the case of online banking, e-commerce or e-voting, the requirements on the degree of security grow enormously.

Cryptology is the science of secret writing and is almost as old as writing itself. In cryptology, we distinguish between **cryptography** and **cryptanalysis**. Cryptography is devoted to the design of cryptosystems (secret codes), while cryptanalysis is devoted to the art of attacking cryptosystems (illegally intercepting messages). At the beginning, when writing was discovered, very few people were able to read or to write. Hence, at that time writing was considered to be secret in the sense that the ability to read was a common secret of a small group of insiders. The need for special secret writing grew with the growing number of people who were able to read. Here we start with providing some impressions about the time when secret writing was an art. One developed a secret code and used it until somebody deciphered it and so made its use insecure. In that time, cryptology was a game between two intellects, namely the secret code designer and a secret code hacker. First, we will have a look at encryptions used by Caesar and Richelieu. In this way, we get the first impression of what cryptosystems are and what the security of cryptosystems means.

Our next aim is to show that, due to computer science, especially due to algorithmics and complexity theory, the art of designing cryptosystems has become a science called cryptography. Cryptography owes a debt to algorithmics for the definition of the notion of the security of cryptosystems. The algorithmic concept of security became the axiomatic base for scientific investigation. Using it, one developed modern cryptography, which is based on the mystery of so-called public-key cryptosystems. Here, we explain the idea behind this mystery that transfers our drawbacks and inability to solve some problems into advantages. The existence of efficiently unsolvable algorithmic problems (which was discussed in Chapter 5) is used to make algorithmic cracking of designed cryptosystems impossible. Finally, we present some miracles that can be viewed as "magic" applications of the concept of public-key cryptosystems.

7.2 The Concept of Cryptosystems or a Few Impressions from the Prehistory of Cryptography

First, we fix our terminology. As already mentioned above, we consider cryptology as teachings of secret writing. We distinguish between cryptography and cryptanalysis. Cryptography is the science of designing secret cryptosystems. Cryptanalysis is devoted to attacking them. Here, we deal with cryptography only. The scenario we consider is presented in Fig. 7.1.

A person, called the **sender**, aims to send a secret message to another person, called the **receiver**. The secret message can be represented as a text, called a **plaintext**. The plaintext can be written in a natural language, and therefore it is readable for everybody. There is no other possibility besides communicating the

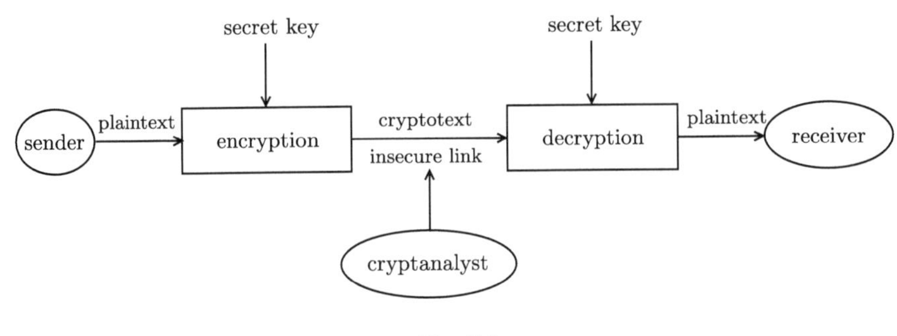

Fig. 7.1

secret through a publicly accessible network, where one cannot exclude the possibility that other persons can "listen" to the communication messages transported in the network. To exclude the possibility that an unauthorized person listening to the message learns about the secret, one sends the message in an **encrypted** form using a secret writing. The kind of **encryption** and of **decryption** is a common secret of the sender and of the receiver, and the encryption is performed using the so-called **key**. The encrypted text is called the **cryptotext**. The cryptotext is sent via the public network. Upon receiving the cryptotext, the receiver **decrypts** the cryptotext and obtains the original plaintext. A description of the processes of encrypting and decrypting the messages provides full information sufficient for using the cryptosystem.

To illustrate it using a transparent example, we present the cryptosystem CAESAR that was really used by Caesar. The plaintext were texts over the Latin alphabets. The special symbols such as blank, comma, period, and question mark were omitted. In this way the text

VENIVIDIVICI

can be viewed as a plaintext. Caesar encoded each letter unambiguously by another letter. The keys are the integers

$$0, 1, 2, 3, 4, 5, \ldots, 23, 24, 25.$$

To apply a key i means that every symbol of the plaintext is replaced by the letter that is k steps (positions) to the right in the alphabetical order. If the end of the alphabet is reached, one continues cyclically from the beginning.

Figure 7.2 shows the encoding using the key $i = 3$. This means that A in the plaintext is represented by D in the cryptotext, because A is at the first position of the Latin alphabet and, shifting by 3 positions, one reaches the fourth position occupied by D. Analogously, the letter B at the second position is replaced by the letter E at the fifth position, etc. In this way, the letter W at the 23rd position is replaced by the letter Z at the 26th position. How to encode the last three letters X, Y, and Z at the positions 24, 25, and 26, respectively? One still did not use the letters A, B, and C in the cryptotext. We simply write the letters A, B, and C at the positions 27, 28, and 29 behind our alphabet, as depicted in Fig. 7.2. In this way, our problem of replacing X, Y, and Z is solved. The letter X is replaced by A; Y is replaced by B; Z is replaced by C. Another transparent way of describing the cryptosystem CAESAR is to write the Latin alphabet on a tape and then stick the ends of the tape together (Fig. 7.3). Now, rotating the tape by 3 positions, one gets the codes of the letters. Applying the key $i = 3$ for the plaintext VENIVIDIVICI, one gets the cryptotext

YHQLYLGLYLFL .

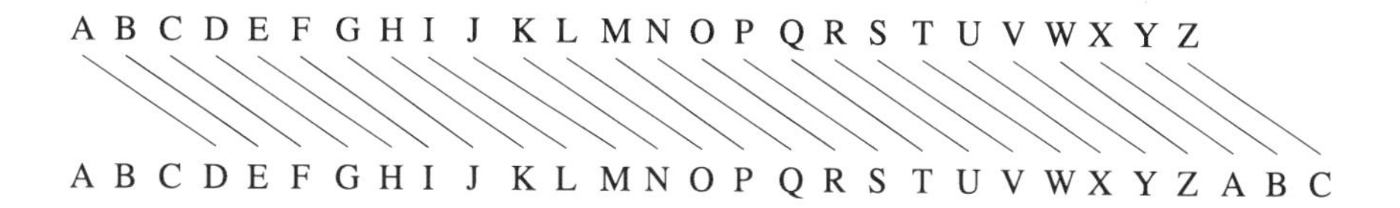

Fig. 7.2

Fig. 7.3

Exercise 7.1 (a) Encode the plaintext VENIVIDIVICI using the key $i = 5$ and then using the key $i = 17$ using the cryptosystem CAESAR.
(b) Encode the plaintext "Roma aeterna docet" using the key $i = 7$, and the plaintext "Homines sumus, non dei!" using the key $i = 13$.

How does the receiver decode the cryptotext? She or he simply replaces each letter of the cryptotext by the letter that is 3 positions before this letter in the alphabet. It corresponds to twisting the tape in Fig. 7.3 three positions to the left. In this way, for instance, the letter S of the cryptotext is replaced by P, U is replaced for R, and L is replaced by I. Hence, one decrypts the cryptotext

SULPXPQHQRFHDV

encrypted using the key $i = 3$ and obtains the original plaintext

PRIMUMNENOCEAS .

Exercise 7.2 Decrypt the cryptotext WYPTBTBAWYVMPJLHZ encrypted using the key $i = 7$.

To be more precise, Caesar tried to decrease the risk that somebody could look through his cryptosystems and therefore he made it more complicated. After shifting the letters, he replaced the letters of the Latin alphabet one by one using the letters of the Greek alphabet. Hence, the receiver had to replace the Greek letters for the Latin ones first, and then to shift positions back inside of the Latin alphabet.

Jules Verne discussed in his novels a cryptosystem generalizing that of Caesar. Any integer was allowed to be a key. For an integer of m digits

$$A = a_1 a_2 \dots a_m$$

he partitioned the plaintext into blocks of length m and replaced the jth letter of each block by the letter that was a_j positions after the letter in an alphabetical order. For instance, for $a = 316$ one gets for the plaintext

C	R	Y	P	T	O	G	R	A	P	H	Y
3	1	6	3	1	6	3	1	6	3	1	6

the cryptotext

FSESUUJSGSIE .

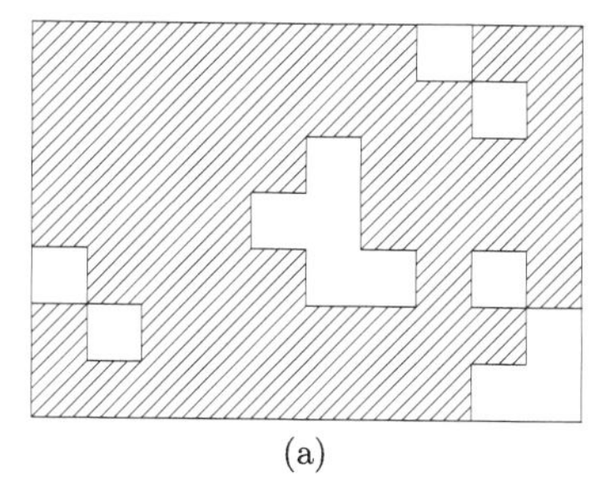

(a)

I		L	O	V	E		Y	O	U
I		H	A	V	E		Y	O	U
D	E	E	P		U	N	D	E	R
M	Y		S	K	I	N		M	Y
L	O	V	E		L	A	S	T	S
F	O	R	E	V	E	R		I	N
H	Y	P	E	R	S	P	A	C	E

(b)

I		L	O	V	E		Y	O	U
I		H	A	V	E		Y	O	U
D	E	E	P		U	N	D	E	R
M	Y		S	K	I	N		M	Y
L	O	V	E		L	A	S	T	S
F	O	R	E	V	E	R		I	N
H	Y	P	E	R	S	P	A	C	E

(c)

Fig. 7.4

In contrast to the factual cryptosystem CAESAR, Richelieu cultivated in his cryptosystem a real literary art. The following example is due to Arto Salomaa [Sal96]. The cryptotext is a well-structured letter on a sheet. Each letter has an exact position on the sheet. For instance, H lies at the intersection of the second row and of the third column in Fig. 7.4(b). The key is the matrix in Fig. 7.4(a) that covers some fields of the sheet (the grey ones in Fig. 7.4(b)) and leaves some fields open for reading. The uncovered letters build the plaintext (Fig. 7.4(c)). For the friendly looking cryptotext in Fig. 7.4(a), the plaintext is

YOUKILLATONCE .

7.3 When Is a Cryptosystem Secure?

People have always strived for security and certainty. The unknown and uncertainty are often the main reasons for fear. In spite of this, science, and especially physics, teaches us that absolute certainty does not exist. To strive for it can even result in an illness. It is better to learn to live with uncertainty. Anyway, it may be reasonable to strive for an achievable degree of security. When is one allowed to consider a cryptosystem to be secure? Always when an adversary of an unauthorized person is unable to get the plaintext from the cryptotext sent. This requirement allows two different interpretations. Does one require it to be hard or even impossible to decrypt a cryptotext if nothing about the used cryptosystem is known or if the cryptosystem (the kind of encryption) is known and only the key is secret? The first option with the whole cryptosystem as a secret is called "security by obscurity", and one does not consider it to be a reasonable base for defining the security of cryptosystems. This is because experience shows that revealing the kind of encryption of a new cryptosystem is only a question of time.

Therefore, already in the 19th century Auguste Kerckhoffs presented the following requirement on the security of cryptosystems, known as **Kerckhoffs' principle:**

> *A cryptosystem is secure, if one, knowing the art of the functioning of the cryptosystem but not knowing the key used, is not able to derive the original plaintext from the given cryptotext.*

We immediately see that under this definition of security the cryptosystem CAESAR is not secure. If one knows the principle of CAESAR, it suffices to try out all 26 keys in order to decrypt any cryptotext.

A direct consequence is that secure systems have to have a huge number of keys. By "huge" we mean that even the fastest computers are unable to try out all of them. The next question is, whether a large number of possible keys guarantees the security of a cryptosystem. Consider an improvement of CAESAR by allowing arbitrary pairs of letters. Then keys are defined by the so-called permutations of 26 letters that can be viewed as 26-tuples of integers between 1 and 26. In any 26-tuple, each integer from the interval 1 to 26 occurs exactly once. For instance, the key

(26, 5, 1, 2, 3, 4, 25, 24, 23, 8, 9, 22, 21, 20, 7, 6, 10, 11, 18, 17, 12, 13, 19, 14, 16, 15)

corresponds to the following encryption. The first letter A is replaced by the 26th letter Z, the second letter B is replaced by the fifth letter E, the third letter C is replaced by the first letter A, etc. The complete description of the resulting pairing of letters is given in Fig. 7.5.

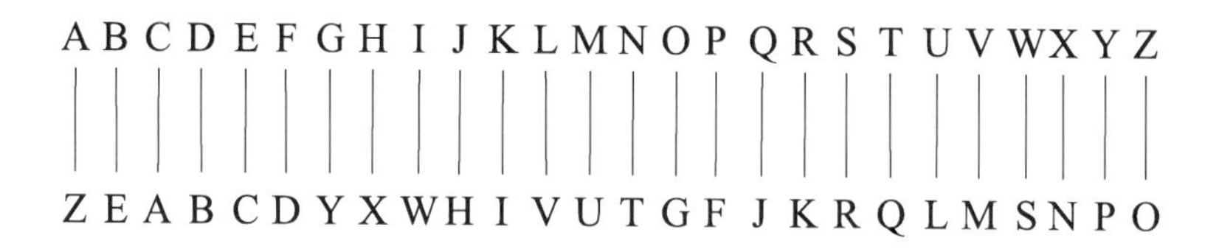

Fig. 7.5

Applying this key, one gets the cryptotext

QXCKCWRTBYKCZQCKBZUZYCQXZTQXCQWUCVGRQ

for the plaintext

THEREISNOGREATERDAMAGETHANTHETIMELOST.

Exercise 7.3 Decrypt the following cryptotexts that were encrypted using the key described in Fig. 7.5.

(a) AKPFQGYKZFXP
(b) RCALKC

How many keys are possible in this cryptosystem? To code A one has the choice from 26 letters of the Latin alphabet. After choosing a letter decoding A, one has 25 possibilities to choose a letter coding B. For C, one has still 24 possibilities, etc. Hence, the number of all keys is

$$26! = 26 \cdot 25 \cdot 24 \cdot 23 \cdot \ldots \cdot 3 \cdot 2 \cdot 1.$$

In the language of combinations, one says that 26! is the number of all permutations of 26 elements. This number is huge, approximately equal to $4.03 \cdot 10^{26}$. No computer is fast enough to be able to try out so many keys.

Does it mean that this system is secure? Unfortunately, “No” is the unambiguous answer. To discover the key applied, one does not need to try out all possibilities. A clever idea (that did not occur to the cryptosystem designer) suffices to crack this cryptosystem. In the case of our cryptosystem based on permutations of letters, it is sufficient to know which natural language is used for the plaintext. For each natural language, one knows the statistical frequency of the occurrence of particular letters, and these frequencies can differ a lot from letter to letter. Additionally, some letter combinations such as TH in English and SCH in German are very frequent. An experienced cryptanalyst can use this knowledge to estimate the key (the letter permutation) and so to decrypt the cryptotext in a few minutes.

This example shows the enormous difficulty of measuring the degree of security (reliability) of a given cryptosystem. What does it mean that it is impossible or hard to crack a cryptosystem? The problem with the definition of security by Kerckhoffs is that one more dimension is involved there. Besides the cryptosystem itself, there is an adversary in the game. What is impossible for one adversary does not need to be hard for another, genius one. An excellent idea may reduce the decryption problem to an easy game. What is one allowed to assume about the facilities of an adversary, or how to get all cryptanalysts under one uniform description?

Computer science solved this long-standing open problem of making the definition of secure cryptosystems clear enough. It covered all simple as well as in ingenious strategies of cryptanalysts using the notion of an algorithm.

A cryptosystem is secure if there does not exist any efficient (randomized polynomial) algorithm that decrypts cryptotext without knowing the secret key used, but knowing the way in which the cryptosystem works.

Clearly, this definition of secure cryptosystems could not be discovered before the notion of algorithm and the concept of computational complexity were introduced. Due to them, one was able to exactly formulate the first two fundamental requirements on a "good" cryptosystem in the scientific terminology provided by complexity theory.

(i) Encryption and decryption can be efficiently performed by algorithms when the key is known.

(ii) Decryption without knowledge of the secret key used in encrypting the plaintext corresponds to a hard (practically unsolvable) problem.

7.4 Symmetric Cryptosystems

The cryptosystems presented until now are called **symmetric**. The term symmetric points to the fact that the sender and the receiver are equal in the sense that the same key as a common secret is used to encrypt as well as to decrypt the messages. In fact, the sender and the receiver can exchange their rules anytime, and so the communication can be viewed as gossiping or as an information exchange.

In what follows, we show a secure symmetric communication system. For this cryptosystem we assume that the plaintexts are simply sequences of zeros and ones. This is not a restriction because each letter of an alphabet as well as each symbol of the keyboard has an ASCII code that is the binary representation of the corresponding letter used in computers. In this way, one can first transform any text into a sequence of zeros and ones.

One also uses as keys sequences of zeros and ones. These sequences have to be long, for instance a few hundred bits, in order to get many keys. They have to be generated at random in order to make their computation impossible for each unauthorized person.

To describe encryption and decryption (i.e., to define the cryptosystem), we have first to introduce a special operation on bits (binary numbers). This operation is called "exclusive or" or "XOR" and is denoted by $\oplus$. It is defined as follows:

$$0 \oplus 0 = 0 \qquad 1 \oplus 1 = 0 \qquad 0 \oplus 1 = 1 \qquad 1 \oplus 0 = 1$$

Following the notation from Chapter 6, one can write $a \oplus b = a + b \bmod 2$. If a and b are equal, the result is 0. If a and b are different, the result is 1.

Assume that the key is 01001011. The length of the key is 8 and the key can be applied to encrypt plaintexts of length 8. The ith bit of the key is used to encode the ith bit of the plaintext. One can see the encryption transparently if one writes the key below the plaintext as follows:

$$\begin{array}{ll} \text{plaintext} & 00001111 \\ \oplus\ \text{key} & 01001011 \\ \hline = \text{cryptotext} & 01000100 \end{array}$$

The ith bit of the cryptotext is the XOR-sum of the ith bit of the plaintext and the ith bit of the key. Interestingly, one can use the same procedure for decryption:

$$\begin{array}{ll} \text{cryptotext} & 01000100 \\ \oplus\ \text{key} & 01001011 \\ \hline = \text{plaintext} & 00001111 \end{array}$$

We see that it works and that the original plaintext was decrypted. The success of this procedure is based on the fact that

$$a \oplus a = 0, \qquad \text{and so} \quad b \oplus a \oplus a = b.$$

Hence, two applications of the same key result in the original text. In another words, the second application of the key deletes the first application. Moreover,

$$a \oplus b = b \oplus a$$

holds, and therefore we call this cryptosystem commutative.

Exercise 7.4 Encode and decrypt the plaintext 0001110011 using the key 0101101001.

Exercise 7.5 Build a similar cryptosystem that is based on the operation $\perp$ defined as follows:

$$0 \perp 0 = 1 \qquad 1 \perp 0 = 0 \qquad 1 \perp 1 = 1 \qquad 0 \perp 1 = 0$$

The second bit b in $a \perp b$ is called a mask. If $b = 1$, then the first bit a is copied. If $b = 0$, then the first bit a flips to $\bar{a}$ (i.e., 1 flips to $\bar{1} = 0$ and 0 flips to $\bar{0} = 1$).

(a) Apply the operation $\perp$ instead of $\oplus$ in order to encrypt and then decrypt the plaintext 00110011 using the key 00101101.
(b) Explain why the cryptosystem based on $\perp$ also works and has the same properties as the cryptosystem based on $\oplus$.
(c) What have the operations $\perp$ and $\oplus$ in common? Does there exist a strong relationship between them?

If the key is generated together by the sender and the receiver[1] at random and then applied in order to mask, one can prove using mathematical instruments that the cryptotext is viewed as a random sequence of bits to any third party. In this case, no cryptoanalyst has a chance to decrypt the message because no algorithm, even ones with exponential complexity, can help. For a single application, this system is considered to be absolutely secure.

If one has a long plaintext and a key of a fixed length n, the usual way of encrypting the text is based on partitioning the text into a sequence of slices of length n and then applying the key separately to each slice. For instance, one can use the key 0101 to encrypt the plaintext

1111001100001101

as follows. One cuts the plaintext into four slices 1111, 0011, 0000, and 1101 and applies the key 0101 to each slice separately. After concatenating the resulting slices, one gets the cryptotext

1010011001011000

[1] That is, the key does not need to be communicated by a public link.

that is sent to the receiver. This is the typical way in which cryptosystems based on a fixed key work for arbitrarily long messages.

In the case of an XOR-cryptosystem, one does not recommend this extension for messages of any arbitrary length, because multiple applications of the same key on several slices can help to calculate the key. Moreover, the fact

$$\text{plaintext} \oplus \text{cryptotext} = \text{key} \tag{7.1}$$

holds. Let us check this property for the key 01001011 used in our example above:

$$\begin{array}{ll} \text{plaintext} & 00001111 \\ \oplus\ \text{cryptotext} & 01000100 \\ \hline =\ \text{key} & 01001011 \end{array}$$

Why is this true?

$$\text{plaintext} \oplus \text{key} = \text{cryptotext} \tag{7.2}$$

is our encryption method. Now, we add the plaintext to both sides of Equality (7.2) from the left and obtain the following equality:

$$\text{plaintext} \oplus \text{plaintext} \oplus \text{key} = \text{plaintext} \oplus \text{cryptotext} \tag{7.3}$$

Since $a \oplus a = 0$ for each bit a,

$$\text{plaintext} \oplus \text{plaintext} = 00\ldots00\ .$$

Since $0 \oplus b = b$ for each bit b, the left side of Equality (7.3) is the key, and hence we obtain Equality (7.1).

The validity of Equality (7.1) can be very dangerous for the security of a cryptosystem. If by a multiple use of a key the adversary obtains only one pair (plaintext, cryptotext) in some way, then by applying (7.1) she or he can immediately calculate the key and so decrypt all other cryptotexts. Therefore, the XOR-cryptosystem is secure for a single application only.

Exercise 7.6 (challenge) Try to design a secure cryptosystem for the following application. A person A encrypts a plaintext (a confidential message) using a secret key that nobody else knows. The confidential message is sent as the cryptotext to two persons B and C. Now, A wants to distribute the secret key to B and C in such a way that neither B nor C alone is able to decrypt any single bit of the message, but B and C cooperating together can calculate the plaintext without any problem.

There are symmetric cryptosystems that are considered to be secure, and one is allowed to use the same key many times without worry. The famous and most frequently used symmetric cryptosystem is DES (Data Encryption Standard) developed in a cooperation between IBM and NSA (National Security Agency). DES also uses the XOR-operation several times, but this cryptosystem is too complex to be described here.

In spite of the high reliability of some symmetric cryptosystems, we are still not finished with trip devoted to searching for a secure cryptosystem. The main drawback of symmetric cryptosystems is that they can be successfully applied only if the sender and the receiver agree on a common secret key before the cryptosystem is applied. But how can they do this without any personal meeting? How can they agree on a secret key without any secure cryptosystem when they are restricted to communication via a public, insecure channel? How to solve this problem is the topic of the next chapter.

7.5 How to Agree on a Secret in Public Gossip?

Two persons, Alice and Bob, want to create a common cryptosystem. The encryption and the decryption procedures are known to both, and they only need to agree on a common secret key. They do not have the opportunity to meet each other and are forced to agree on (to generate) a common secret using a public, insecure channel without applying any cryptosystem. Is this possible?

The answer is "Yes", and the elegant solution is a real surprise. First, we introduce the idea of using a chest with two locks and an untrustworthy messenger. Alice has a private lock with a unique key and so nobody else can close and unlock her lock. Similarly, Bob has his private lock with a unique key. They do not have any common key. Alice and Bob agree publicly that Alice is going to send Bob a future common key in the chest. They proceed as follows:

1. Alice puts the common key into the chest and locks the chest with her private key. Nobody but Alice can unlock the chest because nobody else has a key that fits. Then, she sends the chest to Bob. The messenger cannot unlock the chest.
2. The messenger brings the chest to Bob. Certainly, Bob cannot open it either. Instead of trying to unlock it, Bob locks the chest with his private lock. The chest locked by two independent locks is now sent back to Alice (Fig. 7.6).
3. Alice receives the chest with two locks. She unlocks her lock and removes it. Now, the chest is locked by the private lock of Bob only. Anyway, the messenger as well as nobody except Bob can open it. Alice sends the chest once again to Bob.
4. Now, Bob receives the chest locked by his private key only. He opens the chest and takes the common secret key sent by Alice.

Figure 7.6 taken from [Sal96] shows an amusing graphic outline of the above procedure.

Fig. 7.6

Exercise 7.7 (challenge) Alice aims to send the same common secret key to three other persons using an unreliable messenger in a secure way. One possibility is to repeat the above procedure three times. Doing it, the messenger has to run $3 \cdot 3 = 9$ times between two persons (always between Alice and another person). Is there a possibility to save a run of the messenger?

Can one perform this procedure in an electronic way? First, let us try to do it using the XOR-operation. In this way, one can get a very simple implementation of sending a chest with two locks. As we will see later, this implementation does not fulfill all security requirements. The procedure for an electronic communication is outlined in Fig. 7.7.

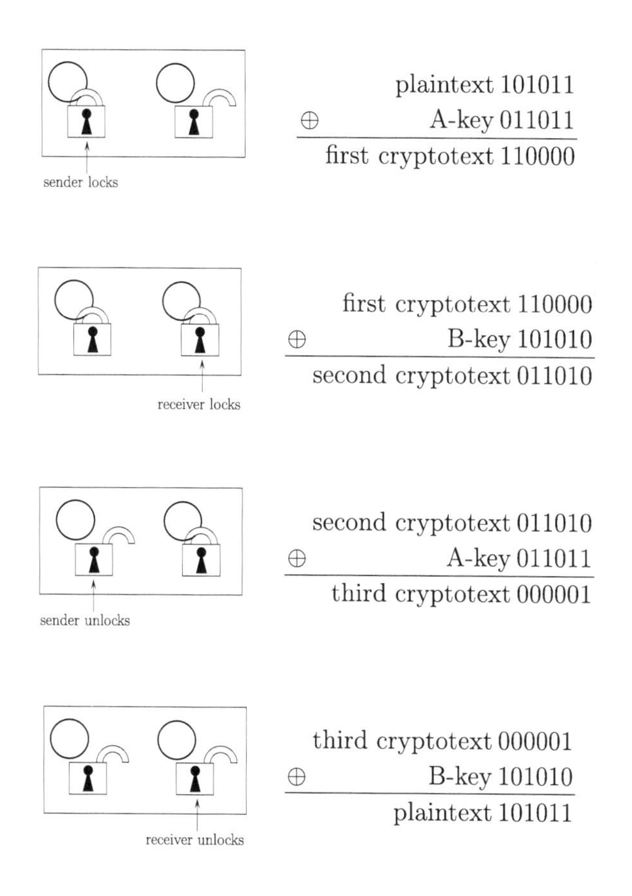

Fig. 7.7

The sender (called Alice) aims to send a common key 101011 to the receiver (Bob). The common key is viewed as the plaintext in Fig. 7.7. Note, the common key Alice and Bob want to agree on as well as their private keys are bit sequences of the same length. As we know, the procedure consists of three communication steps and such procedures are called **communication protocols** in cryptography.

The protocol works as follows:

1. The sender (Alice) with her private key, A-key, calculates

$$\text{plaintext} \oplus \text{A-key} = \text{crypto } 1$$

and sends crypto 1 to the receiver (Bob).

2. The receiver (Bob) with his private key, B-key, calculates

$$\text{crypto } 1 \oplus \text{B-key} = \text{crypto } 2$$

and sends crypto 2 to the sender.

3. The sender calculates

$$\text{crypto } 2 \oplus \text{A-key} = \text{crypto } 3$$

[We observe that

$$\begin{aligned}\text{crypto } 3 &= \text{crypto } 2 \oplus \text{A-key}\\ &= \text{crypto } 1 \oplus \text{B-key} \oplus \text{A-key}\\ &\quad \{\text{because crypto } 2 = \text{crypto } 1 \oplus \text{B-key}\}\\ &= \text{plaintext} \oplus \text{A-key} \oplus \text{B-key} \oplus \text{A-key}\\ &\quad \{\text{because crypto } 1 = \text{plaintext} \oplus \text{A-key}\}\\ &= \text{plaintext} \oplus \text{A-key} \oplus \text{A-key} \oplus \text{B-key}\\ &\quad \{\text{because of the commutativity of } \oplus \text{ one can}\\ &\quad \text{exchange the order of the arguments}\}\\ &= \text{plaintext} \oplus \text{B-key}\\ &\quad \{\text{since } a \oplus a = 0 \text{ and } b \oplus 0 = b\}\end{aligned}$$

holds.]

4. The receiver calculates

$$\begin{aligned}\text{crypto } 3 \oplus \text{B-key} &= \text{plaintext} \oplus \text{B-key} \oplus \text{B-key}\\ &\quad \{\text{since, as shown above,}\\ &\quad \text{crypto } 3 = \text{plaintext} \oplus \text{B-key}\}\\ &= \text{plaintext}\end{aligned}$$

Exercise 7.8 Execute the run of the communication protocol for the key submission for the following data. The plaintext as a future common key is 01001101. The private A-key of Alice is 01010101, and the private B-key of Bob is 10101010.

We explained directly in the description of the communication protocol why the receiver learns the plaintext at the end. The main idea is that applying a key for the second time automatically removes (rubs) the first application of the key, even if there were other actions on the text between these two applications. One can express this idea as

$$\begin{aligned}\text{text} \oplus \text{key} \oplus \text{actions} \oplus \text{key} &= \text{text} \oplus \text{actions} \oplus \underbrace{\mathit{key} \oplus \mathit{key}}_{\text{no action}}\\ &= \text{text} \oplus \text{actions}\end{aligned}$$

Exercise 7.9 The communication protocol above works because the operation $\oplus$ has the nice properties $a \oplus a = 0, a \oplus b = b \oplus a$, and $b \oplus 0 = b$. Investigate whether one can also use the operation $\perp$ to implement this communication protocol.

Is this electronic communication protocol secure? Is its degree of reliability comparable with that of sending the chest? Unfortunately, the answer is no. If a cryptanalyst gets only a particular cryptotext and does not know the rules of this protocol, the communication looks like a sequence of random bits, and the protocol is secure in this sense. Following Kerckhoffs' principle, one has to count on the fact that the adversary knows the rules of our communication protocol. The private keys of Alice and Bob are the only secret for the cryptanalyst. If the adversary reads all three cryptotexts (Fig. 7.7), then he can calculate as follows:

$$\begin{aligned}\text{B-key} &= \text{crypto } 1 \oplus \text{crypto } 2\\ \text{A-key} &= \text{crypto } 2 \oplus \text{crypto } 3\end{aligned}$$

in order to compute the private keys of Alice and Bob.

Exercise 7.10 Use the properties of $\oplus$ to verify the validity of the above equations for calculating A-key and B-key.

After that, the plaintext (the new common secret) is not protected any more, and the adversary can calculate it in any of the following two ways:

$$\text{plaintext} = \text{A-key} \oplus \text{crypto 1}$$
$$\text{plaintext} = \text{B-key} \oplus \text{crypto 3}$$

Hence, this electronic protocol is not considered to be secure. Can one digitize the procedure with the chest and two private locks without losing the degree of security of its "physical" implementation? In 1976, Whitfield Diffie and Martin Hellman [DH76] gave the positive answer to this question. They used the calculator modulo primes introduced in Chapter 6 in a clever way. Without revealing any detailed mathematical argumentation for the correctness of their digital protocol implementation, we describe their procedure in a schema similar to that in Fig. 7.7. Readers who do not like mathematical calculations may omit the following description of the protocol.

The Diffie and Hellman Protocol

Starting point: The sender and the receiver publically agree on two large positive integers c and p, where p is a prime and $c < p$ holds.

The sender (Alice) generates an integer a_A at random. The number a_A is her private key (secret).

The receiver (Bob) generates an integer a_B at random, and this integer is his private key.

The task for Alice and Bob is to calculate a secret in a public communication. This secret is going to be the common key for a symmetric cryptosystem.

Procedure

1. Alice calculates

$$\text{crypto } 1 = c^{a_A} \bmod p$$

and sends crypto 1 to Bob.

2. Bob calculates

$$\text{crypto } 2 = c^{a_B} \bmod p$$

and sends crypto 2 to Alice

3. Alice calculates

$$S_A = (\text{crypto } 2)^{a_A} \bmod p$$

and considers S_A to be the common secret of herself and Bob.

4. Bob calculates

$$S_B = (\text{crypto } 1)^{a_B} \bmod p$$

and considers S_B to be the common secret of himself and Alice.

The kernel of this procedure is that $S_A = S_B$. We omit a mathematical proof of it. But we see that S_A is nothing other than c locked by the key a_B and then by the key a_A. The key S_B is also the encryption of c by the key a_A and then by the key a_B. Hence, both S_A and S_B can be viewed as locking c by a_A and a_B. The only difference between S_A and S_B is that they are locked in different orders. It is the same as locking the chest first by the left lock and then by the right lock, and in the other case first by the right lock and then by the left one. Clearly, for the chest it does not matter in which order the locks were locked. The function $c^a \bmod p$ was chosen here because the order of applying a_A and a_B does not matter either.

If the private keys of Alice and Bob remain secret, then following the correct state of knowledge[2] this digital protocol is secure. But one has to take care when speaking about the security of communication protocols. We clearly defined the meaning of security

[2] We do not know any efficient algorithm that can compute $S_A = S_B$ from c, p, crypto 1, and crypto 2 without knowing any of the secret private keys of Alice and Bob.

for symmetric cryptosystems such as CAESAR or DES. This approach is not sufficient for defining the security of communication protocols exchanging several messages, and so we are required to think once more about it.

The adversary (cryptanalyst) considered until now is called passive. He is allowed to eavesdrop and so to learn the cryptotext and then to try to decrypt it. Diffie–Hellman's protocol is secure against a passive adversary. The situation changes if one considers an **active adversary**, who is allowed to take part in the communication by exchanging messages running in the communication network for other messages calculated by him. Consider the following scenario. The active adversary persuades the messenger to give him the chest, or he removes crypto 1 from the interconnection network. Then, the adversary locks the chest or crypto 1 with his private key and sends it back to Alice. Alice does not foresee that she does not communicate with Bob, but with an adversary. Therefore, following the protocol, she unlocks her lock and sends the chest to Bob. If the messenger once again brings the chest to the adversary, then the adversary can open it and get the secret. Alice is not aware that the secret was revealed.

We see that our protocol is not perfect, and so there is a need to improve it. Whether one can be successful in playing against an active adversary is the topic of the next section.

7.6 Public-Key Cryptosystems

First, let us list the drawbacks of symmetric cryptosystems:

(i) Symmetric cryptosystems are based on a common secret of the sender and the receiver. Therefore, the sender and the receiver have to agree on a fixed key (secret) before using the cryptosystem, hence before having any secure channel for exchanging this secret.

(ii) In reality, one has often a network of agents and a head office accumulating information from all of them. If all agents use the same key, one traitor is sufficient to break the security of the entire secret communication exchange. If each agent has a different key, the center has to manage all the keys, and each agent must reveal his or her identity before sending a message to the head office.

(iii) There is a variety of communication tasks that can be performed using symmetric cryptosystems. In the case of electronic voting (e-voting) one wants, on one side, to convince the system about her or his right to vote, and, on the other hand, to give one's voice anonymously. One needs protocols enabling an authorized person to convince a system about her or his authorization (about possession of a secret or of a password) without revealing any single bit of her or his secret.

These and several other reasons motivated further intensive research in cryptography. When searching for practical solutions again algorithmics and complexity theory provided helpful concepts and instruments. They transformed our drawback concerning the inability to solve hard problems into the strength of cryptosystems. This revolutionary idea is based on the existence of so-called one-way functions. A **one-way** function is a function having the following properties:

(i) The function f is efficiently computable, and so it can be used for an efficient encryption of plaintexts.

(ii) The inverse function f^{-1} that computes the plaintext (argument) x from the cryptotext (value) $f(x)$ (i.e., $f^{-1}(f(x)) = x$) is not efficiently computable. More precisely, there is no efficient randomized algorithm that, for a given cryptotext $=$ f(plaintext), computes the argument $=$ plaintext of f. Hence, nobody can decrypt $f(x)$, and, following our concept of security, the cryptosystem with the encryption function f is secure.

(iii) We must not forget, the authorized receiver needs the possibility of efficiently computing the plaintext from a given cryptotext f(plaintext). There has to exist a secret[3] of f. This secret

[3] called a **trapdoor** in computer science terminology

is something similar to a witness in the concept of randomized algorithms and, knowing it, one can efficiently calculate x from $f(x)$.

What are the advantages of cryptosystems based on a one-way function? The receiver does not need to share the secret with anybody, not even with the sender (Fig. 7.8). The receiver is the only person who possesses the secret, and so the existence of a traitor is excluded. The encryption function f can be publicized, and in this way everybody becomes a potential sender. Therefore, we call cryptosystems based on one-way functions **public-key cryptosytems** (cryptosystems with a public key). Clearly, there is no need for a secure agreement on a common secret before the system starts. In this way, one overcomes the drawbacks (i) and (ii) of symmetric cryptosytems. Moreover, public-key cryptosystems meet the expectation formulated in (iii). But to explain this is beyond the mathematics used in this book.

So far we see that the existence of one-way functions guarantees success in building secure cryptosystems. The main question is: "Do there exist one-way functions?" Are the requirements on a one-way function natural, and do (i) and (ii) not contradict each other? Presenting the following example, we would like to convince you that the concept of a one-way function is not peculiar.

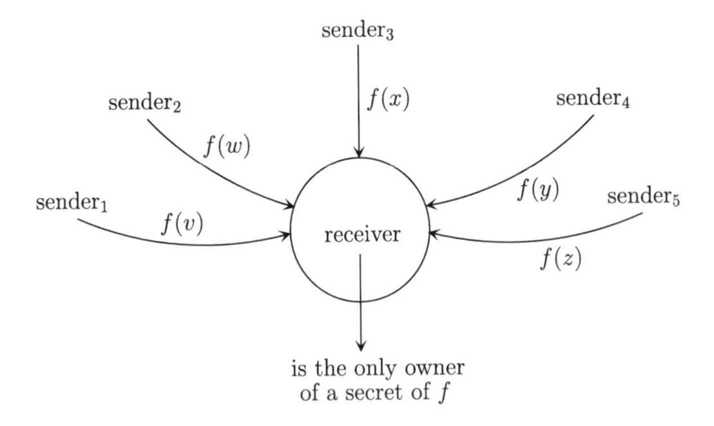

Fig. 7.8

Consider the following encryption. Each letter is separately coded by a sequence of 14 digits. For each letter of the plaintext, one nondeterministically chooses a name from a telephone book that begins with this letter and takes the corresponding telephone number as the encryption. If the number is shorter than 14 digits, one adds the appropriate number of 0's in front of this number. In this way, one can encrypt the word CRYPTOGRAPHY as follows:

	name	telephone number
C	Cook	00128143752946
R	Rivest	00173411020745
Y	Yao	00127345912233
P	Papadimitriou	00372453008122
T	Thomas	00492417738429
O	Ogden	00012739226541
G	Good	00015402316555
R	Rabin	00048327450028
A	Adleman	00173555248001
P	Papert	00016172531555
H	Hopcroft	00013782442358
Y	Yao	00127345912233

Despite the fact that the encryption procedure is nondeterministic and many different cryptotexts may be assigned to a given plaintext, each cryptotext unambiguously determines the original plaintext. The secret (trapdoor) of the receiver is a special telephone dictionary that is sorted with respect to the telephone numbers, and hence the receiver can efficiently execute the decryption. Without knowing this secret (i.e., without possessing the special telephone dictionary sorted with respect to numbers), the work of to decrypt one letter corresponds to the exhaustive search in an unsorted list. This may result in looking at almost all telephone numbers in the dictionary to decrypt one symbol only. Another idea listeners often proposed is to simply try to have a call with all owners of the telephone numbers of the cryptotext. The execution of this idea may result in high costs. Additionally, you have no assurance of learning the name of the telephone owner when calling her or him. If you do not believe it, try this.

Since this encryption method can be published, one can view this cryptosystem as a game on a public-key cryptosystem. We say a "game" because current computers can efficiently sort any telephone book with respect to the telephone numbers, and so create the secret for the decryption. Hence, in reality, one needs other encryption functions.

What are the candidates for one-way functions used in practice? We present three functions that satisfy the requirements (i) and (iii) on one-way functions; we believe the condition (ii) is also satisfied[4].

1. **Multiplication**
   To multiply two primes p and q to compute $f(p,q) = p \cdot q$ can be simply performed by everybody. To calculate the primes p and q from the given value $f(p,q)$ is considered to be a hard problem. The inverse function f^{-1} is related to the factorization problem (to computing prime factors of a given integer), and one knows only exponential algorithms for this problem. For integers the size 500 digits, the fastest computers do not have any chance of finding at least one prime factor.

2. **Modular squaring**
   One can efficiently compute $f_n(x) = x^2 \bmod n$ for any positive public integer n. One simply squares x to x^2 and then divides x^2 by n in order to compute the remainder $x^2 \bmod n$. If n is not a prime (if n is a composite number), one does not know any efficient algorithm that calculates x from the public integer n and a given value $f_n(x)$.

3. **Modular raising to a higher power**
   For public positive integers e and n and a plaintext c in an

[4] To prove that f^{-1} is not efficiently computable is a serious problem, because, as we already mentioned, one lacks mathematical techniques for proving lower bounds on the amount of computer work necessary for computing concrete functions. Researchers have tried to develop such proof methods since the 1960s without any success.

integer representation, one can efficiently compute the value $f_{e,n}(c) = c^e \bmod n$. To show how to do it is a little bit technical and so we do not present it here. If n is not a prime, one does not know any way to compute the plaintext c from the values of e, n, and $f_{e,n}(c)$.

To show what the trapdoors of the above one-way functions are and why the corresponding public-key cryptosystems work requires some knowledge of number theory that is not assumed here. Therefore, we explain only the way to creat a simple public-key cryptosystem based on modular squaring. This system was discovered by Michael O. Rabin, and so it is called RABIN.

Building the cryptosystem RABIN: The receiver generates two large primes p and q at random. The sizes of p and q are about 500 digits. These two primes are the secret (trapdoor) of the receiver. Then, the receiver computes the integer

$$n = p \cdot q$$

and makes n as well as the function $f_n(x) = x^2 \bmod n$ public.

Now, each sender can encrypt her or his plaintext x by computing

$$f_n(x) = x^2 \bmod n$$

and send the cryptotext $f_n(x)$ to the receiver.

Functionality of the cryptosystem RABIN: The senders send their messages $f_n(x) = x^2 \bmod n$ to the receiver. For unknown p and q, no efficient algorithm for computing the plaintext x from given $f_n(x)$ and n is known. But the receiver can efficiently compute x due to her or his secret p and q. The rough idea is the following one. One can efficiently compute x as a modular root of $f_p(x)$ if p is a prime. Number theory allows the receiver to efficiently compute the roots of $f_n(x)$ modulo p and modulo q and then to calculate the plaintext x as a root of $f_n(x)$ modulo $n = p \cdot q$ from these two roots.

Example 7.1 Since we do not want to work with integers of size several hundred digits, we take small primes $p = 107$ and $q = 73$ for the illustration of RABIN. The receiver computes

$$n = p \cdot q = 107 \cdot 73 = 7811$$

and publicizes

> the integer $n = 7811$ and the encryption function $f(x) = x^2 \bmod 7811$.

Now, everybody can encrypt integers smaller than 7811 and send the resulting cryptotext to the receiver.

Assume Alice wants to send the plaintext $x = 6204$. She calculates

$$\begin{aligned} x^2 \bmod n &= (6204)^2 \bmod 7811 \\ &= 38489616 \bmod 7811 \\ &= 4819 \\ &\quad \{\text{since } 38489616 = 7811 \cdot 4927 + 4819\ \} \end{aligned}$$

Hence, the cryptotext is 4819. Since one calculates modulo n, the cryptotext is always an integer smaller than n.

To decrypt the cryptotext, one has to answer the following question. Which numbers from $\{1, 2, \ldots, 7811\}$ squared modulo $n = 7811$ are equal to the cryptotext 4819? Without knowing the factorization $107 \cdot 73$ of 7811, one does not know any better method than to try almost all 7811 candidates from $\{1, 2, \ldots, 7811\}$. For our small $n = 7811$ this is possible. But for big n's such as 10^{1000} it is not physically doable. Knowing the factors of n, one can efficiently estimate all integers y, such that

$$y^2 \bmod n = \text{cryptotext}.$$

At most four such y's may exist. Which of them is the original plaintext can be decided on the basis of the semantics of the text. Another possibility is to ask Alice for an additional bit containing special number theoretic information. □

Exercise 7.11 Build the cryptosystem RABIN for the primes 13 and 17. Then, encrypt the plaintext 100. Try to find all y's from $\{1, 2, 3, \ldots, 13 \cdot 17\}$ with the property

$$y^2 \bmod 13 \cdot 17 = \text{cryptotext}$$

Since we do not consider going deeper into the number theory here, we omit to explain how the receiver applies his secret (the factorization of n) in order to efficiently compute the plaintext. We call attention to the fact that we do not have any mathematical proofs that assure that the presented candidates are really one-way functions. As mentioned above, the trouble is the consequence of the lack of methods for proving the computational hardness of concrete computing problems. Concerning the hardness of the inverse function f^{-1}, the security of all used public-key cryptosystems is only based on the experience that nobody was able to design efficient algorithms for computing f^{-1}. For instance, one knows that cracking RABIN is exactly as hard as factorization of a given integer. More precisely, an efficient algorithm for the decryption of cryptotexts encrypted by RABIN provides an efficient algorithm for factorizing integers, and, vice versa, the existence of an efficient algorithm for factorization implies the existence of an efficient decryption. The size of the randomly chosen primes p and q in RABIN is given by current practice in the sense that the best algorithms for factorization running on the fastest current computers are unable to calculate p and q from a given $n = p \cdot q$ in several billion years. Therefore, the sizes increase with time as algorithms improve and computers become faster.

What are the advantages of public-key cryptosystems?

(i) There is only one secret known to the receiver, and the receiver does not need to share it with anybody else. This secret can be generated by the receiver.

(ii) The procedure of the encryption is publicized. This is the only communication before starting to use public-key cryptosystems and this communication is public, i.e., it does not need any encryption. Everybody can use the public encryption procedure to send encrypted messages to the receiver.

Additionally, public-key cryptosystems can be used for various communication tasks, for which symmetric cryptosystems do not work. To show a simple application, we present a communication protocol for digital signatures. From the judicial point of view, handwritten signature is a form of authenticity guarantee. Obviously, one cannot provide handwritten signatures using electronic communication. Moreover, we would like to have digital signatures that are harder to forge than the handwritten ones.

What do we expect in detail from a communication protocol for a digital signature? Consider the following scenario. A customer K wants to sign an electronic document for her or his bank B. For instance, K has to give the bank B an identity authentication for a money transfer from her or his account. We have the following requirements on communication protocols for such digital signatures:

(i) B must have a possibility of verifying the correctness of the digital signature of K, i.e., to unambiguously authenticate K as the owner of the digital signature. This means that both K and B should be protected against attacks by a third party (a falsifier) F who pretends to be K in a communication with B.

(ii) K must be protected against messages forged by B, who claims to have received them properly assigned from K. Particularly, it means that B cannot be able to forge the signature of K.

(iii) If K signed a document, B has to have a possibility of convincing everybody that the document was signed by K.

Requirement (i) can be achieved also using a symmetric cryptosystem. But no symmetric cryptosystem can satisfy at once the two requirements (i) and (ii).

Exercise 7.12 Design a communication protocol for digital signatures that is based on symmetric cryptosystems and satisfies condition (i).

Satisfying both requirements (i) and (ii) seems to be harder than satisfying the requirement (i) alone, because (i) and (ii) are seemingly contradictory.

> *On the one hand, condition (i) requires that B has some nontrivial knowledge about K's signature for verification purposes. On the other hand, condition (ii) requires that*

B should not know enough about K's signature (or, better to say, about the signature procedure) to be able to forge it.

Surprisingly, using public-key cryptosystems, we can design the following communication protocol that fulfills all three requirements (i), (ii), and (iii).

Construction of the protocol
The customer K has a public-key cryptosystem with a public encryption function E_K and a secret decryption function D_K. The public-key cryptosystem of K is commutative in the sense that

$$E_K(V_K(\text{plaintext})) = \text{plaintext} = V_K(E_K(\text{plaintext}))$$

holds for each plaintext.

What is the profit of this property? Usually, we encrypt a plaintext by computing $E_K(\text{plaintext})$ and then we decrypt the cryptotext using

$$D_K(E_K(\text{plaintext})) = \text{plaintext} .$$

One assumes that the bank knows the public encryption function E_K.

Communication protocol

Input: A document, doc, that has to be signed by K.

Procedure:

1. The customer K computes $D_K(\text{doc})$ and sends the pair

$$(\text{doc}, D_K(\text{doc}))$$

   to the bank B.

2. B computes $E_K(D_K(\text{doc}))$ by applying the public encryption function E_K on the submitted $D_K(\text{doc})$, and checks in this way whether

$$\text{doc} = E_K(D_K(\text{doc}))$$

   holds, and so checks the authenticity of the signature.

Correctness: We call attention to the fact that this electronic signature changes the entire text of the document doc, i.e., the signature is not only an additional text on the end of the document.

Now, we argue that all three requirements are satisfied.

(i) Nobody other than K is able to compute $D_K(\text{doc})$ because only K can do it efficiently. The message $D_K(\text{doc})$ is the only text with the property

$$E_K(D_K(\text{doc})) = \text{doc}$$

and so B is convinced that the message (doc, $D_K(\text{doc})$) was sent by K.

(ii) The knowledge of (doc, $D_K(\text{doc})$) cannot help B to falsify another document w by generating $D_K(w)$ because B cannot efficiently compute the decryption function D_K.

(iii) Since the encryption function E_K is public, the bank B can show the pair (doc, $D_K(\text{doc})$) to everybody interested in it and so everybody can verify the authenticity of K's signature by computing

$$E_K(D_K(\text{doc})) = \text{doc} .$$

This elegant solution to the problem of digital signatures can be viewed as a miracle. But it is only an introduction into the design of several "magic" communication protocols that solve different communication tasks in a surprising way. To understand them, one needs a deeper knowledge of algebra, number theory, and algorithmics, and therefore we refrain from presenting them.

Exercise 7.13 The protocol designed above does not try to handle the document doc as a secret. We see that the customer key makes it public by sending the pair (doc, $D_K(\text{doc})$). Everybody who is listening can learn doc. Now we exchange the requirement (iii) for the following requirement.

(iii′) Nobody listening to the communication between the customer K and the bank B may learn the content of the document doc signed in the protocol.

Design a communication protocol satisfying the conditions (i), (ii) and (iii′).

Exercise 7.14 (challenge) We consider the so-called **authentication problem**. Here, one does not need to sign any document, but one has to convince somebody about her or his identity. The requirements on a communication protocol for the authentication are as follows:

(i′) the same as (i) in the sense that B is convinced about the identity of K, and
(ii′) K should be protected against the activities of B, where B attempts to convince a third party that she or he is K.

The above-presented communication protocol is not satisfactory for authentication because B learns the signature (doc, D_K(doc)) in this protocol and can use it to convince a third party that she or he is K.

Design a communication protocol that satisfies (i′) and (ii′).

We finish this section about public-key cryptosystems with a few important remarks. If one considers that public-key cryptosystems are only useful as one-way communication systems from many senders to one receiver, then one does not see all the possibilities opened by the public-key concept. Everybody can communicate with everybody. Everybody who wants to communicate, generates one's private secret (for instance, p and q) and publicizes the corresponding encryption function (for instance, $n = p \cdot q$) in a public telephone directory. Now, anybody can write to her or him in a secret way by using the published encryption procedure for encrypting the plaintext.

Public-key cryptosystems have many advantages compared with symmetric cryptosystems. As already mentioned, the main advantage is that they are a basis for creating various communication protocols that cannot be built using symmetric cryptosystems. On the other hand, classical symmetric cryptosystems also have an important advantage over public-key ones. Due to possible hardware implementation, symmetric cryptosystems are often hundreds of times faster than public-key ones. Thus, it is common to use a public-key cryptosystem only for exchanging the key of a symmetric cryptosystem. The rest of the communication[5] is then performed using a symmetric cryptosystem.

[5] that is, the main part of the communication

7.7 Milestones of Our Expedition in the Wonderland of Cryptography

Cryptography is devoted to the design of cryptosystems that allow a secure exchange of secret data using public communication channels. At the beginning, cryptography was viewed as the art of creating secret writing. The parties in the communication were the sender, the receiver, and their adversary (the cryptanalyst), and they played an intellectual game. The communicating parties came up with some clever secret codes, and their adversary tried to find tricky ideas enabling her or him to crack the cryptosystem. One was originally unable to define the notion of security in this intellectual game.

Auguste Kerckhoffs formulated the first requirement on the security of a cryptosystem. He insisted that the reliability of a cryptosystem has to be based on the secrecy of the key only and not on the secrecy of the kind of encryption. First, this leads to the idea that a huge number of possible keys is necessary and could be sufficient for building secure cryptosystems. The need for many keys is obvious, but it was quickly recognized that even a huge number of keys does not provide any guarantee for the security of a cryptosystem.

One had to wait for a scientific definition of security, which came when computer science introduced the concepts of algorithms and computational complexity. Using the concept of practical solvability, computer science defined the security of a cryptosystem. A cryptosystem is secure if there is no efficient algorithm transforming a given cryptotext into the corresponding plaintext without knowledge of the key. With this definition, the history of modern cryptography began. Cryptography became a scientific discipline on the border between computer science, mathematics, and, more and more, also of physics.[6]

[6] because of quantum effects used in cryptography. More about this topic will be presented in Chapter 9.

The common property of classical cryptosystems is that one key determines the encryption as well as the decryption, and so the key is a common secret of the sender and of the receiver. The main problem of these cryptosystems is to agree on a secret key before starting the use of the system. We saw that one can design a communication protocol for creating a common secret key. Unfortunately, this communication protocol is secure only against a passive adversary. If the adversary actively takes part in the communication and passes himself off as the receiver in the communication with the sender, then the sender does not detect it and reveals the secret.

One found the way out by designing public-key cryptosystems. Based on the ideas of complexity theory, the concept of so-called one-way functions was proposed. A one-way function can be efficiently computed but its inverse function cannot. The inverse function can be efficiently computed only if one has a special secret called the trapdoor of the inverse function. There is some similarity between the role of the secret for a one-way function and the role of witnesses in the design of randomized algorithms. The one-way function is publicized and used for the encryption of plaintexts. The trapdoor is the receiver's secret and she or he does not need to share it with anybody. Using this secret, the receiver can efficiently decrypt each cryptotext sent to her or him.

Because it is a hard, unsolved problem to develop mathematical methods for proving lower bounds on the amount of computer resources necessary for computing concrete functions, one is unable to provide mathematical evidence that a function is a one-way function. Based on our experience in algorithms, we consider some functions to be one-way functions and use them in practice. Multiplication with factorization as its inverse is a well-accepted candidate for a one-way function. Another candidate for a one-way function is modular squaring, whose inverse function corresponds to computing modular roots of a given integer.

Current applications in e-commerce and in online banking are unthinkable without public-key cryptosystems. Electronic voting is another crucial application area of the concept of public-key cryptosystems.

The idea of public keys was originally proposed by Diffie and Hellman [DH76] in 1976. The most common cryptosystem is the famous RSA cryptosystem designed by Ron Rivest, Adi Shamir, and Leonard Adleman [RSA78] in 1978. Similarly as for other fundamental discoveries, it took 20 years for it to become common in commercial applications.

We recommend the textbooks of Salomaa [Sal96], and Delfs and Knebl [DK07], as involved introduction to cryptography. Solutions to some exercises as well as an extended introduction to this topic can be found in [Hro04b].

Solutions to Some Exercises

Exercise 7.5

(a) We obtain:

$$\begin{array}{r} 00110011 \\ \perp\ 00101101 \\ \hline 11100001 \\ \perp\ 00101101 \\ \hline 00110011 \end{array}$$

(b) This procedure works because

$$(a \perp c) \perp c = a \ .$$

Hence, a double application of a key on a plaintext results in the original plaintext.

(c) The result of applying the operation $\perp$ is always the opposite of the result of applying $\oplus$. More precisely, if $a \oplus b = 1$ holds, then $a \perp b = 0$, and if $a \oplus b = 0$ holds, then $a \perp b = 1$.

Exercise 7.6 Person A generates a sequence $a_1 a_2 \ldots a_n$ of n bits at random. Then A encrypts the plaintext $k_1 k_2 \ldots k_n$ by computing

$$k_1 k_2 \ldots k_n \oplus a_1 a_2 \ldots a_n = d_1 d_2 \ldots d_n \ .$$

After that A generates again n bits $b_1 b_2 \ldots b_n$ at random and calculates

$$\begin{array}{r} a_1 a_2 \ldots a_n \\ \oplus\ b_1 b_2 \ldots b_n \\ \hline c_1 c_2 \ldots c_n \end{array}$$

Finally, A sends the key $b_1b_2\dots b_n$ and the cryptotext $d_1d_2\dots d_n$ to B, and the key $c_1c_2\dots c_n$ and the cryptotext $d_1d_2\dots d_n$ to person C. Since $b_1b_2\dots b_n$ was chosen at random, neither B nor C can estimate the key $a_1a_2\dots a_n$. Without knowing the key $a_1a_2\dots a_n$, there is no possibility to decrypt $d_1d_2\dots d_n$. But if B and C cooperate, they can together compute the key $a_1a_2\dots a_n$ using

$$\begin{array}{l} \quad b_1b_2\dots b_n \\ \underline{\oplus\; c_1c_2\dots c_n} \\ \quad a_1a_2\dots a_n \end{array}$$

Having the original key $a_1a_2\dots a_n$, one can immediately compute the plaintext $k_1k_2\dots k_n = d_1d_2\dots d_n \oplus a_1a_2\dots a_n$.

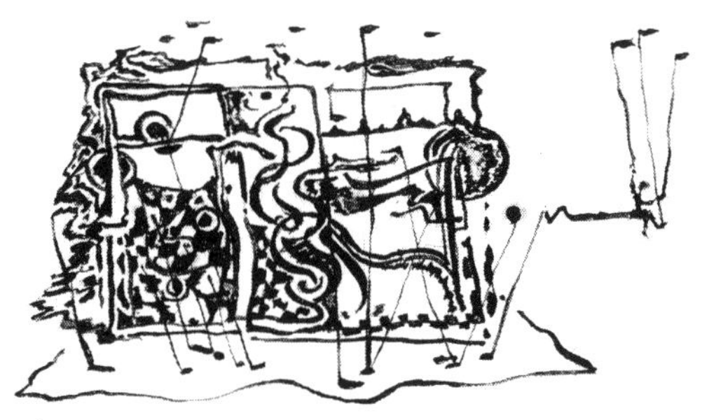

Scientific discoveries are made in the following way.
Everybody knows that something cannot be done.
But then someone comes who does not know about it
and he makes the discovery.

Albert Einstein

Chapter 8

Computing with DNA Molecules, or Biological Computer Technology on the Horizon

8.1 The Story So Far

Many science fiction stories start with a mixture of biological and electronic remainders or connect some parts of the human brain with some computer parts. The final result is then an intelligent robot. The topic of this chapter is removed from such utopian ideas and from the unrealistic promises of some members of the artificial intelligence community in the 1960s. Here we introduce

J. Hromkovič, *Algorithmic Adventures*, DOI 10.1007/978-3-540-85986-4_8,

an existing and non-hypothetical biological information-processing technology. Whether this technology will cause a breakthrough in automatic problem-solving depends on a conceivable progress in the development of biochemical methods for the investigation of DNA sequences.

What is the principle of biocomputers? How did one discover this biological computer technology? Our experience with computer developments in the last 50 years shows that in each period of a few years computers halved in size and doubled in speed an exponential improvement with time. This cannot work forever. Soon, the electronic technology will reach the limits of miniaturization, ending the exponential performance growth of computers. Since the physical limits of the performance of electronic computers were calculated many years ago, the famous physicist Richard Feynman [Fey61] posed already in 1959 the following question: "What is the future of information-processing technologies? Can one miniaturize by performing computations using molecules and particles?". The consequences of these thoughts are the DNA computer and the quantum computer, which will be introduced in the next chapter.

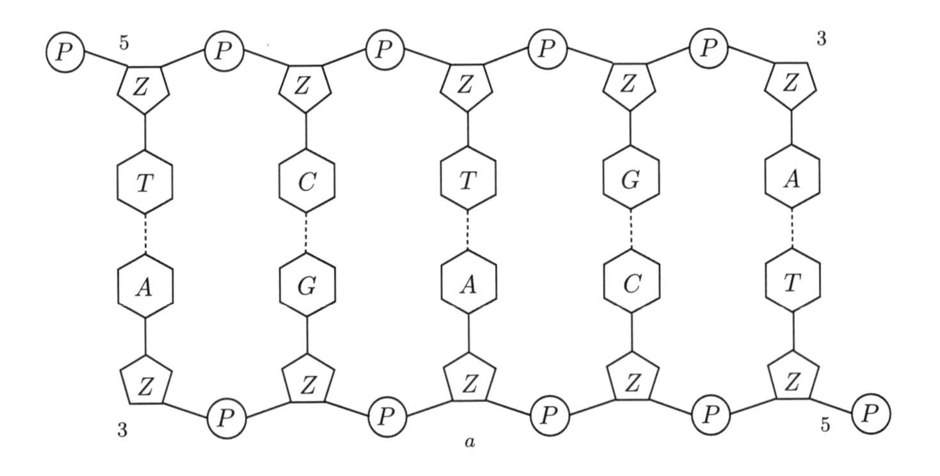

Fig. 8.1

We see that the demand for computer miniaturization and performance growth led to the idea of computing by means of molecules

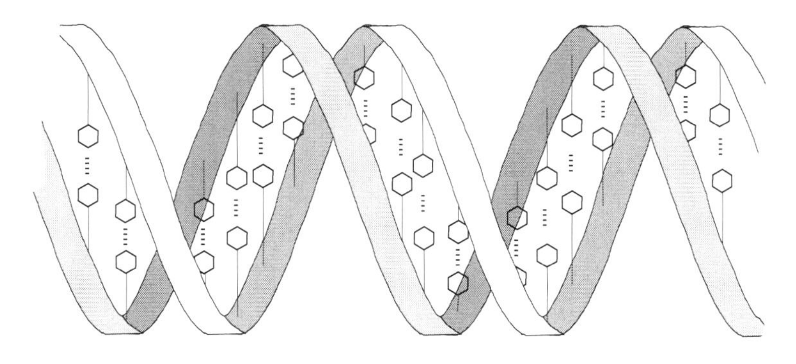

Fig. 8.2

and particles. Is this desire practicable? Is this idea natural? Do we need to force molecules to do work they are probably not built for? To answer these questions, let us first discuss what is natural and what is not natural. We use symbols to calculate in our artificial world of mathematical models. All numbers and all data are represented as texts, i.e., as sequences of symbols. We learned that, in general, the work of a computer can be viewed as a transformation of input texts (input data) into output texts (output data).

What about DNA sequences? We know that they are carriers of biological information and that all processes of living organisms are controlled by the information stored in DNA molecules. Currently, we understand only very little about this control, but there is no doubt about the fact that the biological processes can be viewed as information processing. Unfortunately, because of our restricted knowledge, we are still unable to use this natural information processing to perform calculations of interest. The idea of current DNA computing is much simpler. The DNA sequences can be represented as texts consisting of the symbols A, C, G, and T. These symbols represent the four bases adenine (A), cytosine (C), guanine (G), and thymine (T). Typically, a DNA molecule appears as a double-stranded molecule (Fig. 8.1 and Fig. 8.2). Linkage can be established only between A and T, and G and C. These chemical linkages $\texttt{A}\cdots\texttt{T}$ and $\texttt{G}\cdots\texttt{C}$ are essentially weaker than the other linkages in the string (Z–P in Fig. 8.1). An important fact is that one can build arbitrary DNA sequences, although their stability may vary from molecule to molecule. Hence, we can represent data as texts over the symbols A, C, G, and T, and then DNA sequences can be built as a physical representation of data. After that, one can perform chemical operations with the molecules representing our data and create new DNA sequences. Finally, one can interpret the resulting DNA sequences as the representation of the output.

Surprisingly, one can prove by mathematical means that such a DNA computer can do exactly the same work as classical electronic computers. The consequence is that the notion of algorithmic solvability is not attacked by taking DNA computers into account. All problems algorithmically solvable by an electronic computer can be solved by DNA algorithms, and vice versa.

What is the gain of using a DNA computer instead of an electronic one? A drop of water consists of 10^{19} molecules. If one has 10^{21} DNA molecules in a tube, then a chemical operation with the content of the tube performs in parallel 10^{21} operations on the particular molecules. It is impossible to perform at once one operation over 10^{21} data representations using a classical computer. As a result of this huge parallelism, one can essentially speed up the computer work.

Maybe we are witnessing a new era of a productive concurrency between two different computing technologies. This idea is nicely presented by the pictures (Fig. 8.3, Fig. 8.4, and Fig. 8.5) from the book *DNA Computing* by Păun, Rozenberg, and Salomaa. In Fig. 8.3 one sees the currently standard technology of a classical, electronic computer. In Fig. 8.4 we observe DNA computing performed by chemical operations over the content of tubes in a chemical lab. What can result from this concurrency? Maybe a mixture of electronics and biomass. Instead of performing all chemical operations by hand, one can build an electronic robot performing this task (Fig. 8.5).

In this chapter we first present a list of performable chemical operations with DNA molecules that are sufficient for performing ar-

Fig. 8.3

Fig. 8.4

bitrary computations. After that we present Leonard Adleman's[1] famous experiment. He was the first person to build a DNA computer by solving an instance of the traveling salesman problem in a chemical lab. Finally, we summarize our new knowledge and discuss the drawbacks and the advantages of DNA computing technology. We try to determine the development of biochemical technologies that could ensure a successful future for DNA computing.

[1] Note that Adleman is the same person who was involved in the discovery of the famous RSA cryptosystem.

Fig. 8.5

8.2 How to Transform a Chemical Lab into a DNA Computer

Cooking using recipes and modeling computers in Chapter 2 we saw that to define the notions of "computer" and of "algorithm" one has to fix a list of executable operations and the representation and storage of data. For our model of a DNA computer, data is physically represented and stored by DNA sequences in the form of double-stranded DNA molecules. DNA molecules can be saved in tubes. There are finitely many tubes available.

To perform operations with the contents of tubes, one is allowed to use distinct instruments and devices. We do not aim to explain in detail why one is able to successfully perform the chemical operations considered, because we do not assume that the reader is familiar with the fundamentals of molecular biology. On the other hand, we try to give at least an impression for why some operations are executable. Therefore, we start by giving a short overview of some fundamentals of biochemistry.[2]

[2] An exhaustive introduction for nonspecialists can be found in [BB07, PRS05].

In 1953 James D. Watson and Francis H.C. Crick discovered the double-helix structure (Fig. 8.2) of DNA molecules. The Nobel Prize they got is only a small confirmation of the fact that their discovery counts among the most important scientific contributions of the 20th century. The fact that one can pair only G with C and A with T is called **Watson–Crick complementarity**. A simplified representation[3] of a DNA molecule shown in Fig. 8.1 is the double strand in Fig. 8.6.

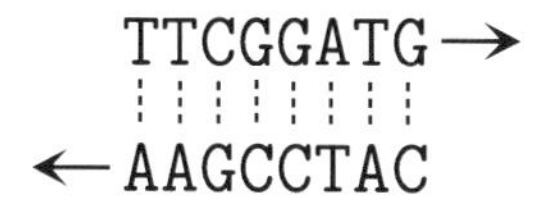

Fig. 8.6

Exercise 8.1 Draw or complete the picture of DNA molecules whose upper strand is AACGTAT, GCCACTA and AACG.

The linkages in the chain

TTCGGATG

are approximately ten times stronger than the chemical linkages (base pairing) A···T and G···C between the bases. The two chains

TTCGGATG

and

AACGCTAC

of the double-stranded molecule in Fig. 8.6 are directed. The upper one goes from left to right and the lower one goes from right to left. The direction is given by the numbering of the carbon atoms of the molecule of the sugar (Fig. 8.7) that connects the phosphate (denoted by P in Fig. 8.1) and the base (A, T, C, or G in Fig. 8.1). There are exactly five carbons in the sugar of a nucleotide[4], as roughly outlined in Fig. 8.7.

[3] DNA molecules have a complex three-dimensional structure that is essential for their functionality.

[4] A nucleotide consists of the phosphate, sugar, and one of the four bases A, T, C, and G.

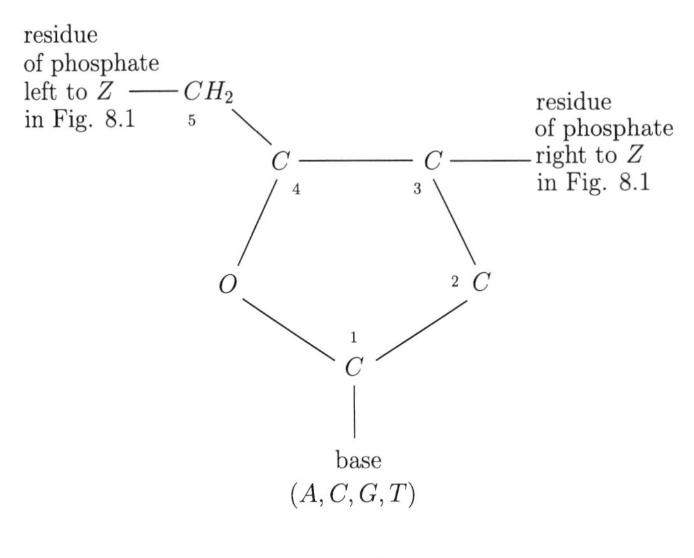

Fig. 8.7

In Fig. 8.7 the molecule of sugar (denoted by Z in Fig. 8.1) is drawn. The 5 carbons C are numbered by $1'$, $2'$, $3'$, $4'$, and $5'$. The carbon $1'$ is responsible for the connection to the base. The carbon $3'$ is responsible for the connection to the phosphate on the right side in the chain and the carbon $5'$ connects the sugar with the phosphate on the left side. In this way, biologists get the direction denoted by $5' \rightarrow 3'$.

Here, it is important to know that by increasing the energy of the DNA molecules (for instance, by heating) one can disconnect the base pairs because the linkages between the pairs A $\cdots$ T and C $\cdots$ G are essentially weaker than the connections inside the strands. In this way, one can split a double-stranded DNA molecule into two single-stranded DNA molecules. Under "suitable" circumstances, two single-stranded DNA molecules may again build together a double-stranded DNA molecule, but only if the base sequences are complementary.

Another important property of DNA molecules is that they are negatively charged. If one measures the length of a double-stranded molecule in the number of base pairs, then one can claim that the size of the charge of a molecule is proportional to its length.

Now, we are ready to introduce some basic chemical operations on the contents of the tubes. Using these operations one can build DNA programs. In what follows, we denote the tubes by R_1, R_2, R_3, etc.[5]

(i) **Union**(R_i, R_j, R_k)
`The contents of the tubes` R_i `and` R_j `are moved to the tube` R_k`.`

(ii) **Amplify**(R_i)
`The number of DNA strands in` R_i `is amplified.`

This operation is based on the Watson–Crick complementarity and is called the polymerase chain reaction. This method revolutionized molecular biology. It was discovered by Kary Mullis in 1985, and for this he was awarded the Nobel prize. First, one heats DNA molecules until the base pairs are disconnected. This allows us to separate the two strands of the double-stranded DNA molecule without breaking the single strands. This phase is called **denaturation**. After that[6], particular nucleotides bind to the single strands in the resulting DNA soup. Since each single-stranded molecule develops to a double-stranded molecule in this way, double-stranded DNA molecules identical to the original ones are created, and the number of DNA molecules in the tube is doubled. This cycle is then repeated several times, and the DNA strands are amplified.

(iii) **Empty?**(R_i)
`One verifies whether there is at least one DNA molecule in` R_i `or whether` R_i `is empty.`

(iv) **Length-Separate**(R_i, l) for an $l \in \mathbb{N}$.
`This operation removes all DNA strands from` R_i `that do not have length` l`.`

To execute this operation, one uses the method called gel electrophoresis, which enables us to measure the length of DNA

[5] We use the same notation as for registers of the memory of a computer model.

[6] This is a simplification of the next phases “priming” and “extension” of the polymerase chain reaction. For more details we recommend [PRS05].

molecules. Since DNA molecules are negatively charged, they move (migrate) towards the positive electrode in an electrical field. The size (length) of the molecule slows down (i.e., the force to move the molecule is proportional to its length) its movement, and the size of its negative charge increases the speed of its movement toward the positive electrode. Since the length of the molecule is proportional to its negative charge, the force of slowing down and the force of speeding up cancel each other. Therefore, the size of the molecule does not matter. Consequently all molecules move to the positive electrode with the same speed. The way out is to put a gel into the electric field that additionally slows down the movement of molecules with respect to their sizes. The consequence is that the small molecules move faster than the larger ones (Fig. 8.8). The speed of movement of the molecules toward the positive electrode can be calculated with respect to their length. If the first molecule reaches the positive electrode, one deactivates the field, which stops the movement of the molecules. Now, one can measure the distances traveled by particular molecules and use them together with the known time of their movement to estimate their lengths. Since the DNA molecules are colorless, one has to mark them. This can be done using a fluorescing substance that intercalates into the double-stranded DNA.

(v) **Concatenate**(R_i)
`The DNA strands in the tube` R_i `can be randomly concatenated.`[7] `In this way some DNA strands longer than the original ones are created.`

(vi) **Separate**(R_i, w) for a tube R_i and a DNA strand w.
`This operation removes all DNA molecules from` R_i `that do not contain` w `as a substring.`

For instance, w = `ATTC` is a substring of x = `AATTCGATC` because w occurs in x as an incoherent part of x. To execute this operation requires a little bit more effort than the previous ones. One possibility is to heat the molecules first in or-

[7] More details about the execution of this operation are given in Section 8.3.

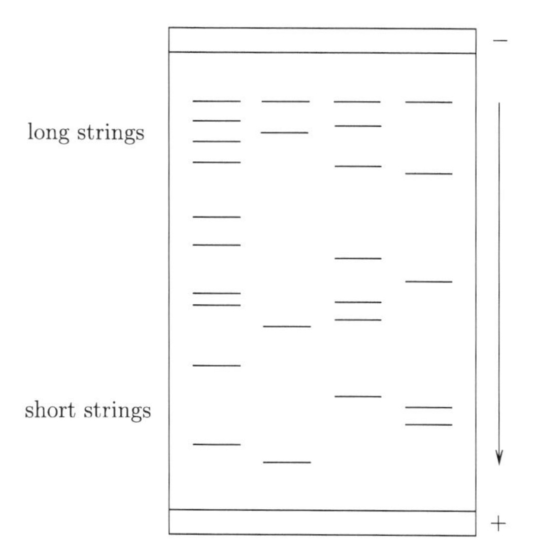

Fig. 8.8

der to split all double-stranded molecules into single-stranded molecules. After that, one puts many strings complementary to w into the tube and cools it down. For instance, for the string w = `ATTC`, the complementary DNA string is `TAAG`. These strings complementary to w will bind themselves on the DNA single-stranded strings containing the substring w. All other strings remain single-stranded. After that one uses a filter that passes only single-stranded molecules through. What remains are strings such as

```
AATTCGATC
 ::::
 TAAG
```

that are neither single stranded nor completely double-stranded. Adding particular nucleotides, one can reproduce complete double-stranded molecules.

(vii) **Separate-Prefix**(R_i, w) for a tube R_i and a DNA strand w.
`All those DNA molecules are removed that do not begin with the string` w.

(viii) **Separate-Suffix**(R_i, u) for a tube R_i and a DNA strand u.
`All those DNA molecules are removed that do not end with the string` u.

Exercise 8.2 Let ATTGCCATGCC, ATATCAGCT, TTGCACGG, AACT, AGCATGCT be the content of a tube R.

Which DNA strands remain in the tube after executing the following operations?

(a) Length-Separate ($R, 7$)
(b) Separate (R, TTGC)
(c) Separate-Prefix (R, TTGC)
(d) Separate-Suffix (R, GCT)

Exercise 8.3 You aim to execute the operation Separate(R, AACT). Which DNA strands have to be put into the tube R after heating in order to be able to select the appropriate ones?

Using this list of eight operations, one can build a DNA computer that is, with respect to computability, exactly as powerful as an electronic computer. How one can solve concrete computing tasks is the topic of the next section.

8.3 Adleman's Experiment, or a Biosearch for a Path

In Section 8.2 we claimed that the chemical operations introduced are powerful enough to simulate the computations of classical computers. We do not aim to provide the evidence of this claim using a mathematical proof. But we do want to give at least an impression of how DNA algorithms can solve classical computing problems. We consider the Hamiltonian Path Problem (HPP) in directed graphs and explain how Adleman solved an instance of this problem in a chemical lab. Adleman's experiment initialized a new interdisciplinary research stream. Recently several leading universities initiated research projects devoted to exploring the possibility of DNA computing technology for solving hard algorithmic problems.

An instance of the HPP is given by a network of roads or air routes. The places (towns) or the road crossings are represented by graph vertices, and the connections between them (Fig. 8.9) are represented by arrows (directed lines). We do not assign any representation to the visual crossing of two lines outside any vertex. Similarly to air routes, there is no possibility to switch between two lines at such crossing points. The directed lines (arrows) are one-way streets. Hence, if one uses the line $Lin(s_1, s_2)$, then after starting at s_1, one must inevitably reach s_2. Another part of the problem instance are the names of two different places s_i and s_j. The task is a decision problem: One has to decide whether it is possible to start at s_i, visit all places of the network exactly once, and finally end at place s_j. A path with the above properties is called a Hamiltonian path from s_i to s_j.

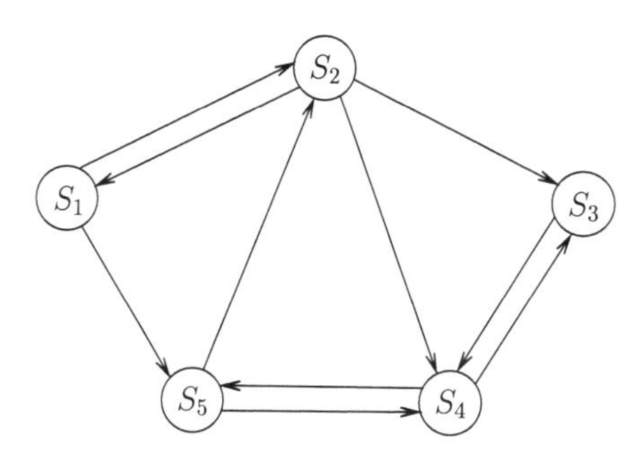

Fig. 8.9

We illustrate this problem using a specific instance. Consider the network depicted in Fig. 8.9 with the starting place s_1 and the goal s_5. Then

$$s_1 \longrightarrow s_2 \longrightarrow s_3 \longrightarrow s_4 \longrightarrow s_5$$

is a solution (i.e., a Hamiltonian path from s_1 to s_5), because each place is visited exactly once on this path. To go along this path one needs the lines $Lin(s_1, s_2)$, $Lin(s_2, s_3)$, $Lin(s_3, s_4)$, and $Lin(s_4, s_5)$. All these four lines are available in the network. Hence, the correct answer to the problem instance in Fig. 8.9 is "YES".

For the network in Fig. 8.9, the starting point s_2 and the goal s_1, there is no Hamiltonian path from s_2 to s_1, and, hence, the correct answer for this instance is “NO”. We see that the destination s_1 can be directly reached only from s_2 by the line $Lin(s_2, s_1)$. But one is required to start AT s_2 and visit all other places before moving to s_1. When one wants to travel to s_1 after visiting all other places, then one is required to go via s_2 to be able to reach s_1. But this is not allowed because one is not allowed to visit s_2 twice.

Exercise 8.4 Consider the network in Fig. 8.10. Does there exist a Hamiltonian path

(a) from s_1 to s_7?
(b) from s_7 to s_1?
(c) from s_4 to s_3?
(d) from s_5 to s_1?

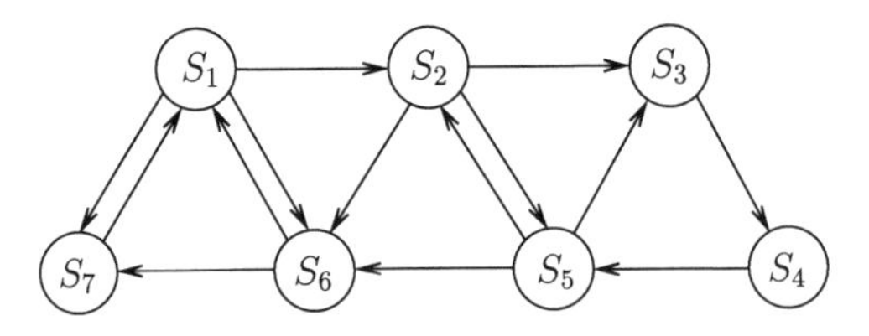

Fig. 8.10

Exercise 8.5 For which pairs of places are there Hamiltonian paths in the network

(a) in Fig. 8.9 ?
(b) in Fig. 8.10 ?
(c) in Fig. 8.11 ?

In what follows we call each sequence of vertices $s_1, s_2, \ldots, s_n$ a path from s_1 to s_n if the directed lines

$$Lin(s_1, s_2), Lin(s_2, s_3), \ldots, Lin(s_{n-1}, s_n)$$

are available in the network. We use the shorter notation $e_{i \to j}$ instead of $Lin(s_i, s_j)$ in what follows. Note that the sequences of vertices s_1, s_7, s_1, s_7, s_1 or $s_7, s_1, s_2, s_5, s_2, s_5, s_6, s_1, s_7$ are also paths because each vertex is allowed to occur several times in a path.

Adleman considered the problem instance depicted in Fig. 8.11 with the starting place s_0 and the destination s_6. His strategy was the following one:

> *Code the name of the places (the vertices) of the network by DNA strands. Then, find a way that enables us to concatenate exactly those place names that are connected by a directed line. Start with as many DNA strings for each place name that by a random concatenation of DNA strings one gets all possible paths of the networks. The code of a path is the corresponding concatenation of the names of the places visited by the path. After that, apply some chemical operations in order to remove all codes of paths that do not correspond to any Hamiltonian path from s_0 to s_6.*

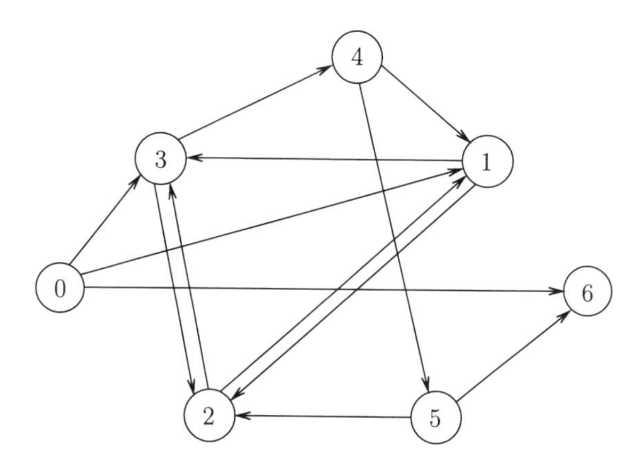

Fig. 8.11

To implement this strategy using chemical operations on DNA strands, he searched for suitable names for the places of the network as sequences of symbols A, C, G, T.

For instance, he chose the following single-stranded DNA molecules for representing s_2, s_3, and s_4:

$$s_2 = \texttt{TATCGGATCGGTATATCCGA}$$
$$s_3 = \texttt{GCTATTCGAGCTTAAAGCTA}$$
$$s_4 = \texttt{GGCTAGGTACCAGCATGCTT}$$

The idea is now to represent directed lines of the network as single-stranded DNA molecules in such a way that using the operation Concatenate(R) only codes of the paths existing in the network can be created.

We can use the property of DNA molecules that only the pairs of bases A, T and G, C can be bound. For each road $e_{i \to j}$ from s_i to s_j we do the following:

- We split the codes of the representations of s_i and s_j in the middle into two equally long strings.
- We create a new string w by concatenating the second part of s_i with the first part of s_j. Then we build the complementary string to w by exchanging A for T, T for A, C for G, and G for C.

Observe that one takes care of the direction of the lines in this way. For our example we get the following codes for the lines $e_{2 \to 3}$, $e_{3 \to 2}$, and $e_{3 \to 4}$:

$$e_{2 \to 3} = \texttt{CATATAGGCT CGATAAGCTC};$$

$$e_{3 \to 2} = \texttt{GAATTTCGAT ATAGCCTAGC};$$

$$e_{3 \to 4} = \texttt{GAATTTCGAT CCGATCCATG}.$$

Figure 8.12 shows how the strings s_2 and s_3 can be concatenated by the line $e_{2 \to 3}$.

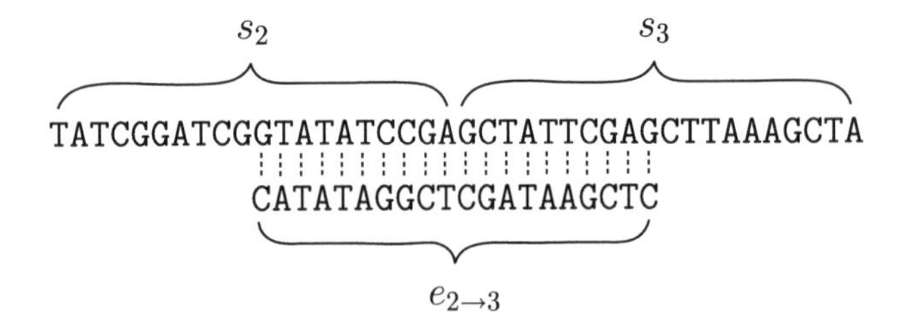

Fig. 8.12

Exercise 8.6 Draw the concatenation of s_3 and s_4 by the line $e_{3\to4}$ in the same way as done in Fig. 8.12 for the concatenation of s_2 and s_3.

Exercise 8.7 Assume one builds a new line $e_{2\to4}$ in the network in Fig. 8.11. Which single-stranded DNA molecule does one have to use to code $e_{2\to4}$? Outline the corresponding concatenation of s_2 and s_4.

If one puts single-stranded DNA molecules coding places and links into a tube, then under suitable circumstances the concatenation of place names (codes) as depicted in Fig. 8.12 and Fig. 8.13 can be performed.

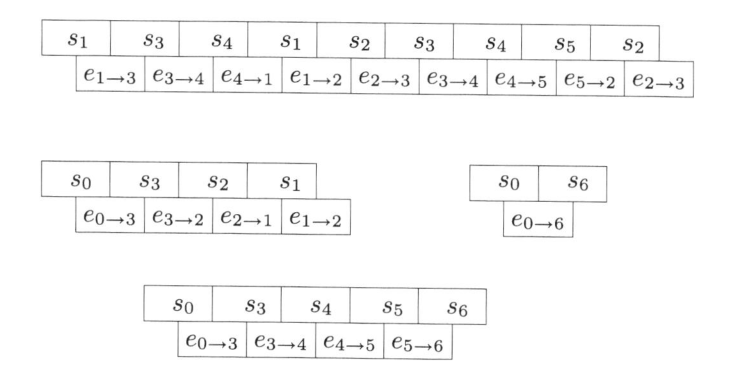

Fig. 8.13

Each created double-stranded DNA string codes a path in the network. To be sure that the process runs exactly in this way, one has to carefully choose the codes of places. The names have to essentially differ each from each other; in particular all halves of the place codes must each differ from each other.

After coding the names of places and streets in the way described above, one can apply the strategy of Adleman, which can be implemented by the following DNA algorithm searching for Hamiltonian paths in a given network of n towns $s_0, s_1, \ldots, s_{n-1}$:

1. Put DNA codes of all places and all roads as single-stranded molecules in a tube T. Add the DNA string of length[8] 10 that is

[8] The length 10 corresponds to our network in Fig. 8.11. Different lengths are possible for the codes of places in other networks.

complementary to the first half of s_0 and a DNA string of length 10 that is complementary to the second half of the destination s_{n-1}.

2. Repeat $(2n \cdot \log_2 n)$ times the operation Amplify(T) in order to get at least n^{2n} copies of each of these single-stranded DNA molecules.
3. Apply Concatenate(T) in order to create double-stranded DNA molecules that code paths in the network of different lengths. The execution of this operation is performed at random. The huge number of DNA codes of the names of places and of links ensures that each possible path of length at most n is generated with high probability.[9]
4. Apply the operation Length-Separate(T, l), where l is n times the length of the code of a place. After executing this operation, tube T contains only DNA codes that represent paths of length n (paths corresponding to sequences of n places).
5. Apply Separate-Prefix(T, s_0). After the execution of this operation, tube T contains only DNA molecules that code paths starting with the code of s_0.
6. Apply Separate-Suffix(T, s_{n-1}) in order to remove all DNA molecules that do not end with the code of s_{n-1}. After that, tube T contains only DNA molecules that code paths of length n that begin with the starting place s_0 and end with the destination s_{n-1}.
7. Apply the sequence of $n - 2$ operations

$$\text{Separate}(T, s_1), \text{Separate}(T, s_2) \ldots, \text{Separate}(T, s_{n-2}) \ .$$

After the execution of this operations sequence, tube T contains only codes of paths that visit at least once each place from $\{s_0, s_1, \ldots, s_{n-1}, s_n\}$. Since the execution of step 4 ensures that all codes correspond to paths of exactly n places, the DNA molecules in T code paths that contain each place exactly once.

[9] In fact, the expectation is that each path is created many times and that paths even longer than n are created.

8. Apply Empty?(T) in order to recognize whether there remained at least one DNA molecule in T. If T is not empty, one gives the answer "YES", else one gives the answer "NO".

The execution of the algorithm for the network in Fig. 8.11 with the starting point s_0 and the destination s_6 results in tube T containing only double-stranded DNA molecules that code the Hamiltonian tour

$$s_0 \longrightarrow s_1 \longrightarrow s_2 \longrightarrow s_3 \longrightarrow s_4 \longrightarrow s_5 \longrightarrow s_6 \; .$$

In this way, one gets the right answer "YES".

Exercise 8.8 Estimate at least three different paths in the network in Fig. 8.11 that remain in the tube after the execution of the fifth operation of Adleman's algorithm. What remains in the tube after the execution of the sixth operation?

Exercise 8.9 Consider the network in Fig. 8.14:

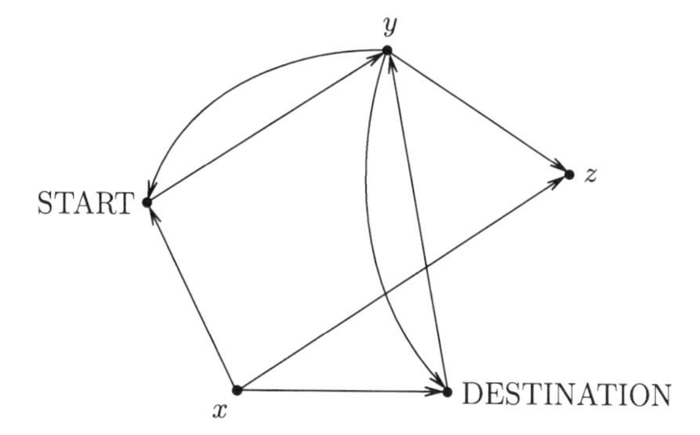

Fig. 8.14

(a) Design codes for the cities in this network as DNA molecules. The codes have to be as short as possible and they have to fulfill the following constraints:
   1. Each code differs in at least four positions from the code of any other city.
   2. The first half of the code of every city differs from the first half of the code of every other city. The second half of the code of each city differs from the code of each other city.

(b) Estimate the codes of the streets of the network in such a way that they correspond to your city codes.

Exercise 8.10 Describe the application of Adleman's algorithm for the problem instance from Exercise 8.9 in such a way that you estimate the values of the parameters such as DNA strings and their lengths.

Exercise 8.11 Is constraint (a) of Exercise 8.9 sufficient to ensure that each path created in the tube corresponds to an existing path of the network? Argue for your answer.

We see that Adleman's algorithm follows a very simple strategy from the algorithmic point of view. It uses the enormous parallelism of DNA computing and creates all possible paths of the network up to some length. All these paths represented by DNA molecules are candidates for a Hamiltonian path. Then, the algorithm deletes all candidates that do not have the properties of a Hamiltonian path from the starting city to the destination.

Since the early 1990s, many DNA algorithms for NP-hard problems were designed and experimentally executed in labs for small problem instances. The current research on this topic deals, on one hand, with increasing the execution time and the reliability of particular chemical operations, and on the other hand with the design of more clever DNA algorithms than those based on the brute force search of all possibilities.

8.4 The Future of DNA Computing

The strength of DNA computing technology lies in the enormous miniaturization that is accompanied by a heavy use of parallelism in data processing. In spite of the fact that the execution of some chemical operations can take several hours or even days, DNA computing can be much faster than classical computing. Due to the high number of DNA molecules in a tube, one could perform computations in a few days that require years to be computed by an electronic computer.

Currently, DNA computers are still no competition for electronic ones, because DNA technology is still in the very beginning of its development. The execution of the operations takes hours and days, and the results are not reliable. We read DNA strands with error rates of 3%, and the majority of the operations may result in creating wrong or unexpected DNA strands. For instance, gaps in the process of getting double-stranded molecules may occur, or wrong strands may be concatenated. Our chemical technologies are simply not able to guarantee correct execution of operations used for DNA computing. To avoid this problem or to decrease the error rate, one uses the concept of redundancy. The idea is to essentially increase the number of copies of the DNA strands as data representations in future and so to increase the probability of getting a correct execution of the operations at least on some strings. The probability of errors grows with the number of executed chemical operations. To compensate for the increase of the error probability, one has to increase the degree of redundancy. The consequence is that for real applications one needs to work with a biomass of the size of the Earth in order to get reliable DNA computations. Hence, skeptics may say, "Let it be, this is only a waste of time. Do something reasonable, something with a foreseeable gain." I would reply that this kind of thinking is the most dangerous type for research management. If one focused in science only on fully foreseeable aims with calculable profit, one would never discover anything essential and we would probably still be climbing trees and know nothing about fire. Let us consider computer technology 40–50 years ago. To place a computer, one needed a large room, and one was required to maintain it daily. We were fighting with cooling systems of permanently warming machines, and we often recognized that something was wrong and consequently all computations of that day had to be repeated. To communicate a designed program to the computer, one used so-called punchcards. Each punchcard was a binary coding (punch or no punch) of one instruction. If somebody tipped over the box containing a long program, it was almost impossible or very awkward to sort the punchcards in the right order. Who believed at that time that computers would become so common that almost everybody uses them? But the development did not take care of the skeptics. Who is able to predict the future of DNA technologies today? Current DNA algorithms would need 18 years to crack the standard cryptosystem DES. If the reliability of the execution of the chemical operations increases, a few hours would be enough.

The future of DNA computing mainly depends on the results of research in molecular biology and chemistry. The degree of precision achievable and the reliability of lab methods are crucial for the competitiveness of DNA computation with respect to classical computation. The possibility of essential improvements cannot be excluded. The duty of scientists is to try to discover the limits of this technology. I believe that several fascinating "miracles" are waiting for us in these research areas.

Adleman's vision about using DNA molecules goes beyond the horizon we see today. Information processing cannot be reduced to the activities of a computer or of our brains. Information processing is running in biological systems and physical systems as a natural phenomenon. DNA molecules contain information for controlling processes for the production of some proteins and even for controlling mechanisms for the choice of the appropriate DNA program. Does this mean that a particular molecule can behave as a programmable computer? If one understands the programs represented by DNA substrings and the corresponding biological mechanisms, then one can open up the possibility of programming a molecule as a universal computer. In this case, nobody will need tons of biomass for getting reliable information processing. Our algorithms would be programmable as parts of one molecule.

For non-biologists interested in the fundamentals of the manipulation of DNA molecules we recommend the book by Böckenhauer and Bongartz [BB07]. The textbook by Drlica [Drl92] is an excellent introduction to molecular biology. An exhaustive and enthusiastic introduction to DNA computing is presented in [PRS05]. A short description of this concept is also given in [Hro04a].

If you do not expect the unexpected,
you will never find jewels,
which are so difficult to seek.

Heraclitus

Chapter 9

Quantum Computers, or Computing in the Wonderland of Particles

9.1 Prehistory

Physics is a wonderful science. If I had had a good physics teacher in high school, then I would probably have become a physicist. But I do not regret that I became a computer scientist. If one reaches the true depth of a particular scientific discipline, then one unavoidably touches also other areas of basic research. In this way, one gains insight into the fundamentals that are common for all sciences, and so sees many things clearer and finds them more exciting than people looking at everything from the specific

J. Hromkovič, *Algorithmic Adventures*, DOI 10.1007/978-3-540-85986-4_9,

angle of their own scientific discipline.[1] Physics provides simultaneously a deep and a wide view on the world. Particularly in the nineteenth century and in the first half of the twentieth century no other science shaped our world view more than physics. Exciting discoveries, unexpected turns, and spectacular results were everyday occurrences in physics. For me, quantum mechanics is one of the greatest achievements of human creativity. Understanding and believing in quantum physics was no less difficult than relinquishing the belief in the Middle Ages that the Earth occupied the central position in the Universe. Why did quantum mechanics have similar troubles with acceptance as did the work of Galileo Galilei? The laws of quantum mechanics provide the rules for the behavior of particles and these rules are in contradiction with our experience about the behavior of objects in our macro world. In what follows, we list the most important principles of quantum mechanics that broke the world view of classical physics:

- An object is located at any moment[2] in one place. This does not hold for particles. For instance, an electron can be in several places at the same time.
- The principle of causality says that each action has unambiguous consequences, or that any cause has an unambiguously determined effect. This is not true for particles. In some scenarios chance conquers. One cannot predict the effects of some actions. There exist several possibilities for the further development of the current situation, and one of these possibilities is chosen at random. There is no possibility to calculate and so to predict which of the possibilities will be executed. One can only calculate the probability of the execution of the particular possibilities. Because of this, physicists speak about true random events in the corresponding scenarios.

[1] Too strong a specialization and the resulting narrow point of view are the sources of the main difficulties of current science, that, on the one hand protect researchers from real success, and on the other hand give rise to intolerance and disdain between scientists from different areas of science.

[2] According to the theory of relativity, time is also a subjective (relative) notion.

- The principle of locality says that effects are always local. In the wonderland of particles, two particles may be so strongly connected to each other that, independently of their distance from each other (maybe even billions of light years), a change in the state of one of the particles simultaneously causes a change in the state of the other particle.
- Classical physics says that if an event is possible (i.e., if an event has a positive probability of occurrence), then it will occur with the corresponding frequency. This is not necessarily true in the wonderland of particles. Two events that may occur with positive probability can completely cancel each other out, as waves can. The consequence is that none of these events can occur.

It is a surprise that physicists were able to discover such strange laws and even make them plausible. How did they do it? They did it in the common way that physicists used since the existence of their science. Researchers developed ingenious ideas by experiments, observations, and related discussions, and were able to express their concepts using the language of mathematics. Using mathematical models, physicists were able to make predictions. If all calculated predictions were confirmed by experiments, one had good reason to believe in the trustworthiness of the models. The experimental confirmation of the theory of quantum mechanics took many years, because an essential development in experimental physics was necessary to reach this purpose. Moreover, the development of experimental equipment was often very costly.

What are we trying to tell you here? Reading a few books is not sufficient to get a good understanding of quantum mechanics. The related mathematics is not simple and the degree of hardness additionally increases when one tries to design quantum algorithms for solving tasks. Therefore, we are satisfied with presenting somewhat fuzzy ideas on how particles behave and how to use this behavior for algorithmic information processing.

In the next section, we first visit the wonderland of particles, observe their behavior in some experiments, and try to explain it.

In Section 9.3 we show how one can store bits in quantum systems and how to calculate with them. We also discuss concepts and problems related to building a quantum computer. Section 9.4 closes this chapter by discussing the perspectives of quantum computing for solving hard computing problems and for the development of secure cryptosystems.

9.2 A Short Walk in the Wonderland of Quantum Mechanics

Alice was in Wonderland only in her dreams. Physicists had it harder than Alice. Doing their job, they were in the wonderland of particles every day and were not allowed to wake up and run away. They were forced to do their best in order to explain at least partially the astonishing behavior of particles. The hardest point was that they were additionally required to update the recent view on the functioning of our world. To say goodbye to one's ideas and viewpoints is not simple and this process takes some time. A still harder job is to convince other scientists that the old theories are not completely true and that the theories have to be revised by using strange concepts that contradict our experience to date. Physicists have to feel lucky that the validity of physical theories is not a subject for a referendum. How to proceed in order to achieve at least a partial acceptance and understanding by the people living in the macro world of the anomalous world of particles? Let us start by investigating some experiments.

First, we consider the famous double-slit experiment (Figs. 9.1, 9.2, 9.3). We have a source of light used to shoot photons in all possible directions. In front of this light source there is a solid plate (Fig. 9.1) that has two small slits. One can open or close the slits like windows. One or both of the slits may be open. Behind the solid plate, there is a photographic plate (the bold line in Fig. 9.1) that registers the photon's hit. In Fig. 9.1 we observe the situation in which the left slit is open and the right slit is closed. The horizontally drawn curve in Fig. 9.1 shows the frequency of

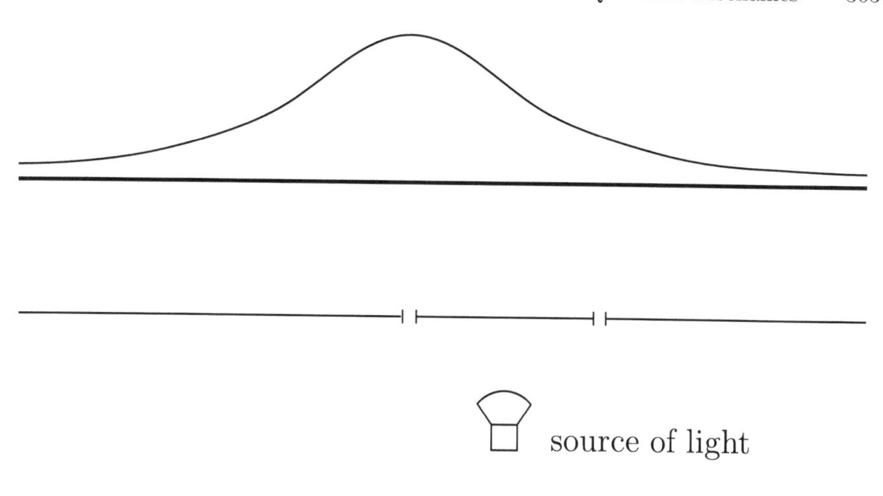

Fig. 9.1

photons coming to the corresponding part of the photographic plate. This curve corresponds to our expectation. Most particles reach the area in front of the slit, and the frequency of incoming particles decreases with the distance from the position in front of the slot. As expected, we get the same picture in Fig. 9.2, when the left slit is closed and the right slit is open. The frequency and so the probability of reaching a concrete place on the photographic plate increases as we approach the place in the front of the right slit.

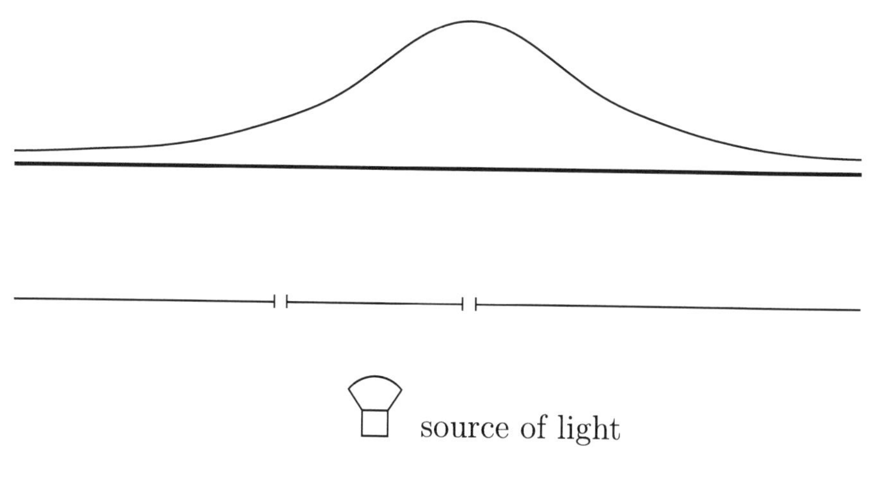

Fig. 9.2

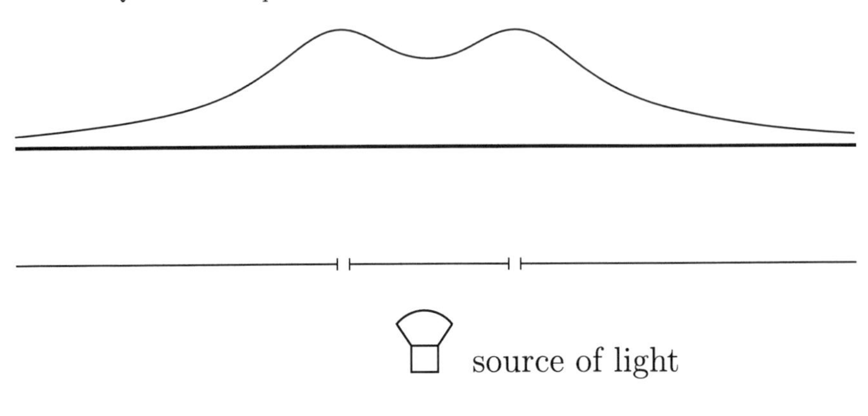

Fig. 9.3

If both slits are open, then one could expect the resulting frequency to be the sum of the frequencies obtained by one open slit and one closed slit in Fig. 9.1 and Fig. 9.2. The expected curve is drawn in Fig. 9.3. Surprisingly, the observed frequency of photons approaching the photographic plate in the experiment of two open slits is completely different, as one can see in Fig. 9.4.

The frequency curve drawn in Fig. 9.4 is very far from the expected curve in Fig. 9.3. Nevertheless, physicists do not view the curve in Fig. 9.4 as an unknown one, or even as a chaotic one. They immediately see that it corresponds to wave interference. If one starts two waves in the slits, the waves will mutually erase each other at some places, and strengthen at other places. In this context, one speaks about wave interference. The result of erasing and strengthening waves corresponds exactly to the frequency curve in Fig. 9.4. How can we explain this? The photons (or other particles) are shot sequentially one after the other, and so two different photons cannot interfere one with the other. The only explanation is that each photon passes simultaneously through both slits and interferes with itself. The kernel of this experiments provides a basic idea of quantum computing. A particle is to some extent in the left slit and partially also in the second slit. If the occurrence of the photon in the left slit represents the bit value 0 and the occurrence of the photon in the right slit represents 1, then the resulting value is something in between 0 and 1. Not a number between 0 and 1:

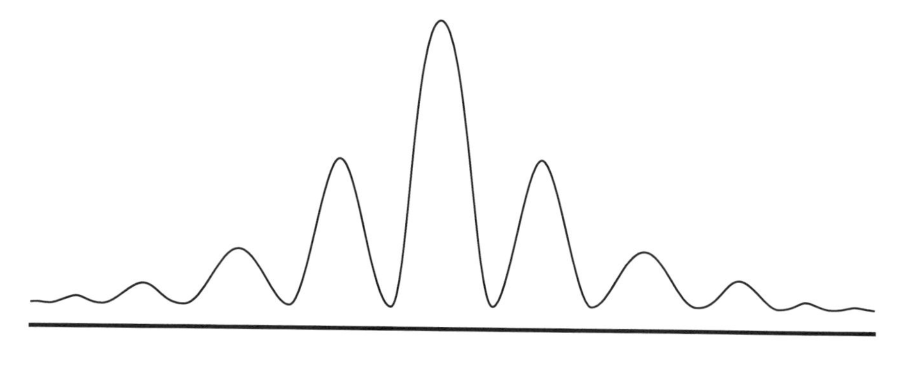

Fig. 9.4

The bit value is partially 0 and partially 1. This is something that cannot happen in the macro world. The mathematical model of quantum mechanics is based on this idea that a particle is allowed to be at several places at one time, and so is able to interfere with itself. The predictions calculated by this model completely agree with the result of verifying experiments (Fig. 9.4).

This is not the end of the story. The experimenter can decide to carefully follow the behavior of the photon. She or he uses another light source on the side in order to detect which of the two slits is used by particular photons. Surprisingly, she or he observes that each photon uses exactly one of the slits to cross the metal plate, and no photon uses both at once. But the frequency curve changes to the expected curve in Fig. 9.3.

The particles are really clever specimens. They behave as goody-goody persons in a village. If one observes them, they do exactly what one expects from them. If nobody looks, then they do incredible things. One can switch between observing and not observing the results beyond the slits as many times as one likes, the frequency curves change corresponding to Fig. 9.3 and Fig. 9.4. If one reduces the power of the observer light in such a way that only a part of the photons can be observed, then the resulting frequency curve is a mix of the curves in Fig. 9.3 and Fig. 9.4. The more light, the more similarity to the curve in Fig. 9.3. The more darkness, the greater the resemblance to the curve in Fig. 9.4. How to explain this behavior of photons?

Quantum mechanics provides an answer. First, one has to imagine the following general law:

> *It is impossible to make an observation or a measurement of a physical object without influencing the state of the observed object and consequently the values of the measurement.*

Exactly that happens in our experiment. The additional light source of the observer influences the result of the experiment by fixing the observed photon in one of the slits. Each observation of a quantum system results in the collapse of the system into a so-called **classical state**. A classical state corresponds to our classical world. A particle is here or there, but never at several places at once. Whether our observation of the double-slit experiment fixes the photon in the left slit or in the right one cannot be calculated and we cannot influence it. It happens randomly according to the probabilistic laws of quantum mechanics. In the next section, we will see that this fact is crucial for the development of quantum algorithms.

If you are uncomfortable due to the above presentation of the double-slit experiment, do not take it too seriously. Physicists needed several years to learn to live with the concept of quantum mechanics. More precisely, a generational change was necessary to accept quantum phenomena. Since I hope that the reader who survived reading the book up till now is strong enough for further surprises, I allow myself to present one more physical experiment.

The following experiment shows the enormous power and importance of interference. Consider the following holiday idyll. The sun is shining, there are no clouds in the sky, there is no wind, and we see the blue sea. A friendly fish is swimming a few meters below starting from the sun. The fish is disturbed in its heavenly peace only by a beam of light that reaches one of its eyes. Which of the possible ways does the light beam take to come from the sun to the eye of the fish? Following physical laws, we know that the light takes the time-shortest path from the sun to the fish eye (Fig. 9.5), and that this path does not correspond to the shortest path between the sun and the eye (the broken line in Fig. 9.5). Since light is faster in the air than in water, the light rays take a longer path in the air into account in order to shorten the path in the water below the sea surface. Therefore, the light beam changes its direction when reaching the sea surface (the solid line in Fig. 9.5).

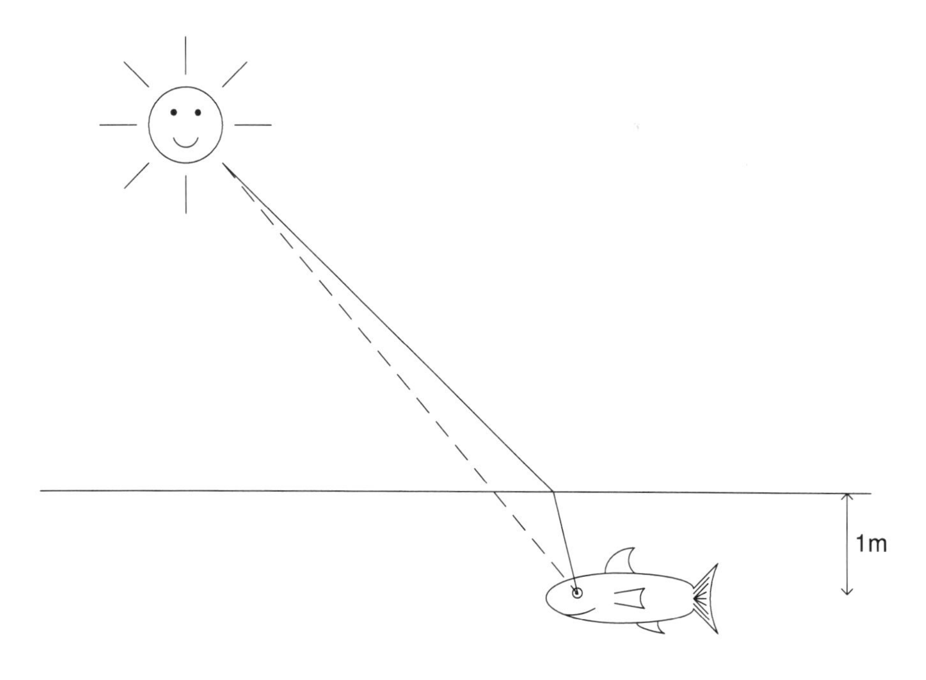

Fig. 9.5

If one wishes, one can calculate the corresponding angle. Now, one can think that the light ray is always refracted by a fixed angle. But take care, let us consider another fish that swims 100 m below the sea surface (Fig. 9.6). In this case, the path of the light ray in the water is shortened more than in the previous case. Therefore, the time-shortest path reaches the sea surface to the right of the place the light coming to the first fish reaches the surface, and refracts by a larger angle (Fig. 9.6) than the first light ray in Fig. 9.5.

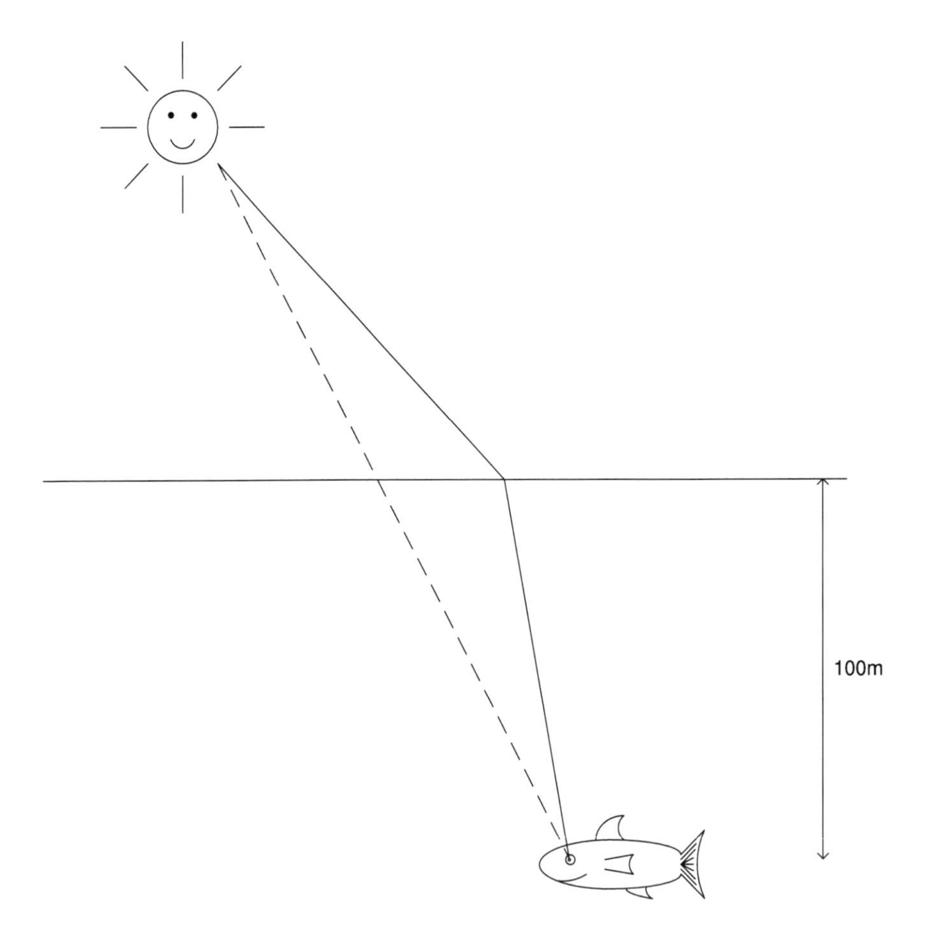

Fig. 9.6

Following Fig. 9.5 and Fig. 9.6, we see that the angles the light rays refract are different. Is it possible? How can the light ray starting from the sun know which of the two fishes it wants to go to and then change its direction correspondingly? How can a photon starting from the sun decide which fish it wants to reach and calculate the time-shortest path to the object chosen? Photons seem to be very clever. Classical physics can observe that the light always takes the time-shortest path from the source to the goal, but it cannot explain how this is performed. However, quantum mechanics can. Obviously, the light beams cannot calculate and plan their trajectory. The photons simply run (radiate) in all directions and try to reach the fish in all possible ways. But all the possibilities that do not correspond to the time-shortest path interact as waves and erase each other. Only the photon taking the time-shortest path survives. What can one learn from this experiment? There is a calculation. But no photon performs the calculation itself. The calculation is done by the interference of the photons and follows the laws of quantum mechanics. These and similar calculations are permanently performed in nature. We cannot observe them. We can only see their results. How we can use such calculations for solving algorithmic problems is the topic of the next section.

9.3 How to Compute in the World of Particles?

In Chapter 2 we saw the basic concept for building a computer. Roughly speaking, we have a memory for storing data and the ability to process (change) the data using some operations. The use of a classical computer assumes storing data as bit sequences and using arithmetic operations to work with data. In the case of cooking, the memory consists of different containers, and the hardware for executing operations are devices of different kinds such as a microwave, cooker, mixer, etc. A DNA computer saves data as DNA sequences in tubes and executes chemical operations over the DNA sequences in a tube. How can we build a quantum computer? We see that we first need to fix the way in which data are represented and stored, and then we have to describe operations over the data representation chosen.

To build a quantum computer, we use bits to represent data in the same way as in the classical computer. We also work with registers that are called **quantum registers** here. A quantum register can store one **quantum bit**. To distinguish quantum bits from classical ones, 0 and 1, one uses the notation

$$|0\rangle \text{ and } |1\rangle \ .$$

There are several ways to build physical quantum registers. The next three short paragraphs are devoted to readers interested in physics. It is possible to skip this part without losing the ability to read the rest of this chapter.

One possibility for building quantum bits is based on nuclear magnetic resonance. Figure 9.7 shows how four of the six atoms of a molecule can be used as quantum registers. If the molecule is in a magnetic field, the spin of the nucleus becomes directed parallel to the field. One can interpret this parallel direction of spins as $|0\rangle$. The direction vertical to the field direction is then associated with the quantum bit $|1\rangle$ (Fig. 9.7). One can use oscillating fields to execute operations on these quantum bits.

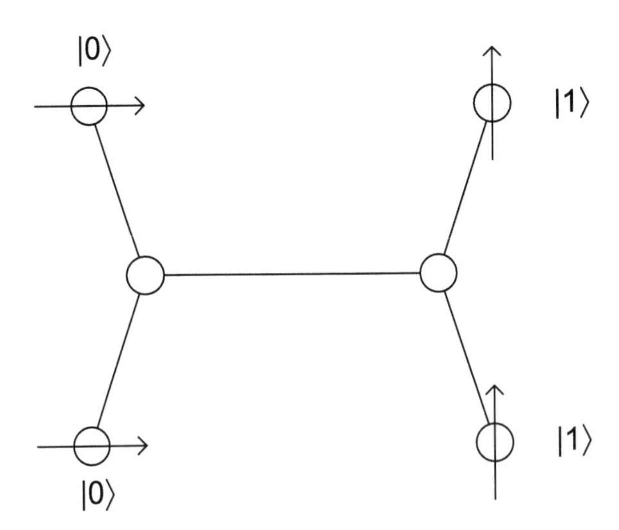

Fig. 9.7

Another possibility is the use of ionic traps. Ions are electrically charged molecules or atoms. In Fig. 9.8, the ions are positively charged, because each one is lacking two electrons. The ions are held in vacuum at a temperature close to absolute zero and at an electromagnetic field in an ionic trap.

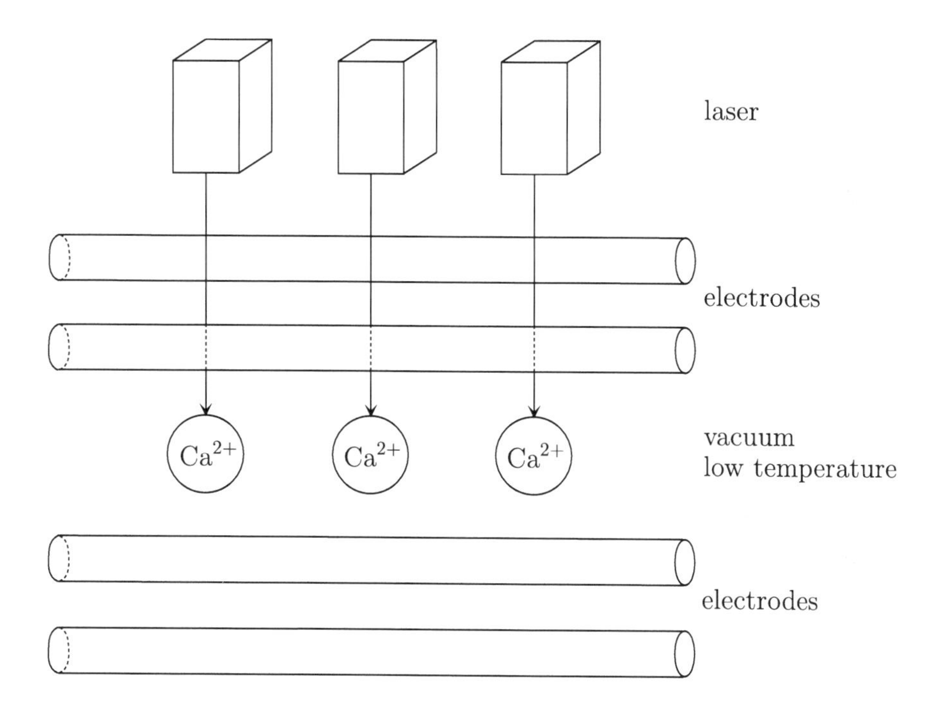

Fig. 9.8

The quantum value $|0\rangle$ is assigned to the basic state of the ion, and the value $|1\rangle$ is represented by an energetically excited state of the ion. Quantum operations over particular quantum bits can be performed using laser beams.

How does a quantum computer compute, and what are its advantages? Considering the bit register of a classical computer, then this register always contains either 0 or 1. This is not the case for a quantum computer.

A quantum register can simultaneously contain both possible contents, each one to some extent. This is the same situation as in the double-slit experiment, where a photon passes through both slits at once; to some extent through the left slit and to some extent through the right slit. How shall we describe it?

One says that a quantum bit (as the contents of a quantum register) can be a **superposition** (or in a combination) of two classical bits $|0\rangle$ and $|1\rangle$. We describe superposition by

$$\alpha \cdot |0\rangle \quad + \quad \beta \cdot |1\rangle \quad ,$$

where α and β are complex numbers satisfying the properties

$$|\alpha|^2 \leq 1, |\beta|^2 \leq 1 \text{ and } |\alpha|^2 + |\beta|^2 = 1 \; .$$

The notation $|\alpha|$ represents the norm of α. If you do not know what complex numbers are, you do not have to be irritated. You can read the remainder of the chapter without any problem, because we use only real α and β in what follows. For any real number α, $|\alpha| = \alpha$, and so our conditions on α and β are simplified to

$$\alpha^2 \leq 1, \beta^2 \leq 1 \text{ and } \alpha^2 + \beta^2 = 1 \; .$$

The values α and β are called **amplitudes**, and they estimate to which extent the quantum bit is $|0\rangle$ and to which extent it is $|1\rangle$.

The exact interpretation is as follows:

α^2 is the probability that $|0\rangle$ is the contents of the quantum register.

β^2 is the probability that $|1\rangle$ is the contents of the quantum register.

The constraint $\alpha^2 + \beta^2 = 1$ is a consequence of this interpretation, because there is no other possibility for classical states than $|0\rangle$ and $|1\rangle$, and the sum of all probabilities of a random register is in the state (superposition)

$$\alpha \cdot |0\rangle + \beta \cdot |1\rangle \; .$$

We do not have any possibility to completely learn this superposition, i.e., there is no possibility to measure both α and β and so to learn their values. If one performs a measurement of a quantum bit, then one sees one of the classical values $|0\rangle$ or $|1\rangle$ only. The measurement definitely destroys the original superposition. This is the same situation as we have in the double-slit experiment. A photon goes through both slits, but if one observes the slits, then the photon goes through one of the slits only.

In our interpretation, α^2 is the probability that the result of the measurement is $|0\rangle$. Analogously, the classical bit $|1\rangle$ is the outcome of the measurement with the probability β^2.

Example 9.1 The superposition

$$\frac{1}{\sqrt{2}} \cdot |0\rangle + \frac{1}{\sqrt{2}} \cdot |1\rangle$$

expresses the fact that the quantum bit has the same probability of being in the classical state $|0\rangle$ as in $|1\rangle$.

$$\alpha^2 = \left(\frac{1}{\sqrt{2}}\right)^2 = \frac{1}{(\sqrt{2})^2} = \frac{1}{2} \quad \text{and} \quad \beta^2 = \left(\frac{1}{\sqrt{2}}\right)^2 = \frac{1}{2} \quad .$$

Because of this when repeating the measurement of this superposition many times, we see $|0\rangle$ as often as $|1\rangle$. □

Example 9.2 The superposition

$$\frac{1}{\sqrt{3}} \cdot |0\rangle + \sqrt{\frac{2}{3}} \cdot |1\rangle$$

expresses the fact that after a measurement, one sees the outcome $|0\rangle$ with the probability

$$\alpha^2 = \left(\frac{1}{\sqrt{3}}\right)^2 = \frac{1}{3}$$

and $|1\rangle$ with the probability

$$\beta^2 = \left(\sqrt{\frac{2}{3}}\right)^2 = \frac{2}{3} \, .$$

□

Exercise 9.1 Propose a superposition of one quantum bit, such that one measures $|1\rangle$ with probability $\frac{1}{4}$ and one measures $|0\rangle$ with the probability $\frac{3}{4}$.

How does this work in the general case? If one has n quantum registers, then one has 2^n possible classical contents. A superposition of n quantum bits means that the register is at once in all 2^n possible classical states, in each one with some probability. The only constraint is that the sum of these 2^n probabilities must be 1.

Example 9.3 Consider two quantum registers. All their possible contents are the following four ones:

$$00, \quad 01, \quad 10, \quad \text{and} \quad 11.$$

In the case of two quantum registers, the contents of the memory of the quantum computer is in superposition

$$\alpha \cdot |00\rangle + \beta \cdot |01\rangle + \gamma \cdot |10\rangle + \delta \cdot |11\rangle$$

with

$$\alpha^2 \leq 1, \beta^2 \leq 1, \gamma^2 \leq 1, \delta^2 \leq 1 \text{ and } \alpha^2 + \beta^2 + \gamma^2 + \delta^2 = 1 \quad .$$

The concrete superposition

$$\frac{1}{2} \cdot |00\rangle + \frac{1}{2} \cdot |01\rangle + \frac{1}{2} \cdot |10\rangle + \frac{1}{2} \cdot |11\rangle$$

with $\alpha = \beta = \gamma = \delta = \frac{1}{2}$ describes the situation where all classical possible contents have the same probability

$$\alpha^2 = \beta^2 = \gamma^2 = \delta^2 = \left(\frac{1}{2}\right)^2 = \frac{1}{4}$$

of being measured.

Consider the superposition

$$0 \cdot |00\rangle + 0 \cdot |01\rangle + 0 \cdot |10\rangle + 1 \cdot |11\rangle \quad .$$

Since $\alpha = \beta = \gamma = 0$ and $\delta^2 = 1^2 = 1$, every measurement has the classical outcome

$$|11\rangle \quad .$$

For the superposition

$$\frac{1}{\sqrt{2}} \cdot |00\rangle + 0 \cdot |01\rangle + 0 \cdot |10\rangle + \frac{1}{\sqrt{2}} \cdot |11\rangle$$

one can measure only the outcomes $|00\rangle$ or $|11\rangle$, both with the same probability $\frac{1}{2}$. □

Exercise 9.2 How can we give a general description of the superposition of three quantum bits?

(a) Give the superposition of three quantum bits such that all possible classical contents of three registers have the same probability to be the outcome of a measurement.
(b) Find a superposition of three quantum bits, for which the classical contents $|111\rangle$ is measured with probability $\frac{1}{2}$, the value $|000\rangle$ with probability $\frac{1}{4}$, and all other possible contents with the same probability.

What operations are possible in the world of particles? How to transform a superposition in a quantum computing step into another superposition? All computation steps are possible that can be performed by a multiplication of superpositions (represented by vectors) by a matrix from a special class of matrices. These matrices have the special property of being able to create a new superposition from each given superposition.

The rest of this section is devoted only to readers with an interest in mathematics. All others may jump to Section 9.4.

A superposition

$$\alpha_1 \cdot |00\rangle + \alpha_2 \cdot |01\rangle + \alpha_3 \cdot |10\rangle + \alpha_4 \cdot |11\rangle$$

can be represented by the column vector

$$\begin{pmatrix} \alpha_1 \\ \alpha_2 \\ \alpha_3 \\ \alpha_4 \end{pmatrix} .$$

If one wants to write a column vector in a row, then one writes $(\alpha_1, \alpha_2, \alpha_3, \alpha_4)^{tr}$, where tr stands for transposed. One can multiply a row vector with a column vector as follows:

$$(\beta_1, \beta_2, \beta_3, \beta_4) \cdot \begin{pmatrix} \alpha_1 \\ \alpha_2 \\ \alpha_3 \\ \alpha_4 \end{pmatrix} = \alpha_1\beta_1 + \alpha_2\beta_2 + \alpha_3\beta_3 + \alpha_4\beta_4 \quad .$$

The result is a complex number, not a vector. One can view an $n \times n$-matrix as n row vectors. For instance, the 4×4 matrix

$$M = \begin{pmatrix} a_{11} & a_{12} & a_{13} & a_{14} \\ a_{21} & a_{22} & a_{23} & a_{24} \\ a_{31} & a_{32} & a_{33} & a_{34} \\ a_{41} & a_{42} & a_{43} & a_{44} \end{pmatrix}$$

consists of the following four row vectors:

$$\begin{gathered} (a_{11}, a_{12}, a_{13}, a_{14}) \\ (a_{21}, a_{22}, a_{23}, a_{24}) \\ (a_{31}, a_{32}, a_{33}, a_{34}) \\ (a_{41}, a_{42}, a_{43}, a_{44}) \end{gathered}$$

Hence, the multiplication of the matrix M by the column vector $\alpha = (\alpha_1, \alpha_2, \alpha_3, \alpha_4)^{tr}$ is again a column vector $\mu = (\mu_1, \mu_2, \mu_3, \mu_4)^{tr}$. The i-th position of μ is the product of the i-th row vector of M and α. More precisely,

$$M \cdot \begin{pmatrix} \alpha_1 \\ \alpha_2 \\ \alpha_3 \\ \alpha_4 \end{pmatrix} = \begin{pmatrix} \mu_1 \\ \mu_2 \\ \mu_3 \\ \mu_4 \end{pmatrix} \quad ,$$

where

$$\begin{aligned} \mu_i &= (a_{i,1}, a_{i,2}, a_{i,3}, a_{i,4}) \cdot \begin{pmatrix} \alpha_1 \\ \alpha_2 \\ \alpha_3 \\ \alpha_4 \end{pmatrix} \\ &= a_{i,1} \cdot \alpha_1 + a_{i,2} \cdot \alpha_2 + a_{i,3} \cdot \alpha_3 + a_{i,4} \cdot \alpha_4 \quad . \end{aligned}$$

The application of M on a superposition is considered one computation step. If for each superposition α (represented as a column vector), the product $M \cdot \alpha$ is also a superposition (in our example it means that $\mu_1^2 + \mu_2^2 + \mu_3^2 + \mu_4^2 = 1$), then multiplication by M is allowed as a computing operation.

In the following example, we show how a quantum computer can generate random bits.

Example 9.4 We have one quantum register. We start with the "classical" superposition

$$|0\rangle = 1 \cdot |0\rangle + 0 \cdot |1\rangle \ .$$

We perform one computing step by multiplying this superposition with the Hadamard matrix:

$$H_2 = \begin{pmatrix} \frac{1}{\sqrt{2}} & \frac{1}{\sqrt{2}} \\ \frac{1}{\sqrt{2}} & -\frac{1}{\sqrt{2}} \end{pmatrix} .$$

The result of the multiplication is as follows:

$$\begin{pmatrix} \frac{1}{\sqrt{2}} & \frac{1}{\sqrt{2}} \\ \frac{1}{\sqrt{2}} & -\frac{1}{\sqrt{2}} \end{pmatrix} \cdot \begin{pmatrix} 1 \\ 0 \end{pmatrix} = \begin{pmatrix} 1 \cdot \frac{1}{\sqrt{2}} + 0 \cdot \frac{1}{\sqrt{2}} \\ 1 \cdot \frac{1}{\sqrt{2}} + 0 \cdot \left(-\frac{1}{\sqrt{2}}\right) \end{pmatrix} = \begin{pmatrix} \frac{1}{\sqrt{2}} \\ \frac{1}{\sqrt{2}} \end{pmatrix} \ .$$

In this way we get the superposition

$$\frac{1}{\sqrt{2}} \cdot |0\rangle + \frac{1}{\sqrt{2}} \cdot |1\rangle \ .$$

If one performs a measurement on this superposition, then one has the same probability of getting the classical bit $|0\rangle$ as of getting $|1\rangle$.

If one starts with the "classical" superposition

$$|1\rangle = 0 \cdot |0\rangle + 1 \cdot |1\rangle$$

and again multiplies our superposition by the matrix H_2, one gets

$$\begin{pmatrix} \frac{1}{\sqrt{2}} & \frac{1}{\sqrt{2}} \\ \frac{1}{\sqrt{2}} & -\frac{1}{\sqrt{2}} \end{pmatrix} \cdot \begin{pmatrix} 0 \\ 1 \end{pmatrix} = \begin{pmatrix} 0 \cdot \frac{1}{\sqrt{2}} + 1 \cdot \frac{1}{\sqrt{2}} \\ 0 \cdot \frac{1}{\sqrt{2}} + 1 \cdot \left(-\frac{1}{\sqrt{2}}\right) \end{pmatrix} = \begin{pmatrix} \frac{1}{\sqrt{2}} \\ -\frac{1}{\sqrt{2}} \end{pmatrix} .$$

The result is the superposition

$$\frac{1}{\sqrt{2}} \cdot |0\rangle - \frac{1}{\sqrt{2}} \cdot |1\rangle .$$

Since $\alpha^2 = \left(\frac{1}{\sqrt{2}}\right)^2 = \frac{1}{2}$ and $\beta^2 = \left(-\frac{1}{\sqrt{2}}\right)^2 = \frac{1}{2}$, the outcomes $|0\rangle$ and $|1\rangle$ of a measurement have the same probability $\frac{1}{2}$. What does it mean? In both cases, we obtain a random bit, but we are unable to distinguish which of the two superpositions $\frac{1}{\sqrt{2}} \cdot |0\rangle + \frac{1}{\sqrt{2}} \cdot |1\rangle$ and $\frac{1}{\sqrt{2}} \cdot |0\rangle - \frac{1}{\sqrt{2}} \cdot |1\rangle$ was measured. □

Exercise 9.3 Give at least two additional different superpositions of a quantum bit, such that the outcomes $|0\rangle$ and $|1\rangle$ are measured with the same probability.

Exercise 9.4 (challenge) Prove that H_2 has the property that

$$\begin{pmatrix} \gamma \\ \delta \end{pmatrix} := H_2 \cdot \begin{pmatrix} \alpha \\ \beta \end{pmatrix}$$

for each superposition $\alpha \cdot |0\rangle + \beta \cdot |1\rangle$ the resulting column vector $(\gamma, \delta)^{tr}$ also represents a superposition, i.e., that $\gamma^2 + \delta^2 = 1$.

Exercise 9.5 (challenge) All computations of a quantum computer are reversible. If one does not measure and so does not destroy the achieved superposition, then it is possible to let the computation run backwards to the starting point using any feasible computing steps. For Example 9.4 this means that there exists a 2×2 matrix M, such that

$$M \cdot \begin{pmatrix} \frac{1}{\sqrt{2}} \\ \frac{1}{\sqrt{2}} \end{pmatrix} = \begin{pmatrix} 1 \\ 0 \end{pmatrix} \quad \text{and} \quad M \cdot \begin{pmatrix} \frac{1}{\sqrt{2}} \\ -\frac{1}{\sqrt{2}} \end{pmatrix} = \begin{pmatrix} 0 \\ 1 \end{pmatrix} .$$

Find a matrix M with the above property.

Now, we understand at least to some extent how one can compute in the world of particles. First, one has to cleverly assign the values 0 and 1 to some different basic states of particles (for instance, the spin direction). After that, one can perform quantum operations with the superpositions of quantum bits. Mathematically, one executes an operation by multiplying the vector representations of superpositions with a matrix. Only those matrices are allowed that transform superpositions to superpositions. All operations based on such matrices can be performed in the quantum world. Here, we see the advantages of quantum computers. If one computes with n quantum bits, then one has a superposition of all 2^n possible classical contents of the n quantum bits:

$$\alpha_0 \cdot |00\ldots0\rangle + \alpha_1 \cdot |00\ldots01\rangle + \ldots + \alpha_{2^n-1} \cdot |11\ldots1\rangle \,.$$

The whole superposition is changed in one simple computing step. To simulate one computing step of a quantum computer using a classical computer, one cannot do better than to multiply the 2^n-dimensional vector $(\alpha_0, \alpha_1, \ldots, \alpha_{2^n-1})^{tr}$ with a $2^n \times 2^n$ matrix, i.e., than to exactly follow the mathematical model of quantum mechanics. In this way, the effort to simulate one single quantum computation step is exponential in the number of quantum bits.

Another strength of quantum computing is the already mentioned interference that enables us to erase existing possibilities and simultaneously to increase the possibility of the occurrence of other possibilities using suitable quantum operations.

Currently, we know several problems for which quantum algorithms are essentially more efficient than the best known classical algorithms. Unfortunately, the knowledge of mathematics and algorithmics necessary for presenting them is too complex to show here. Therefore, we only mention, for instance, that there exists an efficient quantum algorithm for the factorization of natural numbers. As we already know we are unable to efficiently factorize numbers using classical algorithms. This is dangerous for public-key cryptography that is based on the computational hardness of factorization. But, do not worry, you may continue to sleep well. The largest quantum computers built have at most seven quantum bits, and so are only able to work with numbers that can be represented using seven bits. As we know, in public-key cryptography one works with numbers consisting of several hundred digits. Hence, the future will decide about the usefulness of quantum computing for real data processing. This is the topic of the next section.

9.4 The Future of Quantum Computing

Currently, we do not know the future of quantum computing. To understand the difficulty of developing quantum computing technology, let us explain where there are serious troubles. Using the mathematical model of quantum processes, we discovered a tool for designing quantum algorithms. For some problems, these quantum algorithms can be unexpectedly efficient. To execute them, one needs a quantum computer with as many quantum bits as the size of the data to be processed. We also have an idea of how to physically represent the states of particles used as quantum registers. What is missing? A quantum computer is extremely sensitive. It is more sensitive than everything you know in the classical world. To tell somebody "you are more sensitive than a quantum computer" is already a serious insult. The penetration of a single particle such as an electron into a quantum computer in use can result in a complete change of the superpositions computed until that point. We are unable to reconstruct this superposition, and so we must start the computation over from the very beginning. One particle penetrating into the quantum computer can have the same consequences as a measurement. Therefore, it is necessary to completely isolate quantum computers from their surroundings. But this task is harder than to build a secret treasury bond. Particles are everywhere, even in the substance one would use for isolation in the classical world. This is the reason for using vacuums, temperatures close to absolute zero, etc. One is aware that it is impossible to isolate a system from its environment forever. The task is only to isolate the quantum computer for a fraction of one second, because the quantum computation can be performed very quickly. The basic idea is not to construct a general quantum PC that could be used for a variety of different applications. The aim is to build a quantum computer that can solve only one specific computing task. This means that for performing one quantum algorithm designed, one has to build a quantum computer that does nothing other than execute this single algorithm. Hence, we speak about one-purpose (or special-purpose) computers that can solve only one specific computing task. For a small number of bits (3 to 7), physicists have been able to execute some quantum algorithms using small quantum computers. To some extent, the situation is similar to DNA computing. Currently, nobody can consider this technology as a serious competition or addition to classical computers. But when one discovers a better technology for building quantum computers, the situation will change dramatically. We will be required to revise the definition of the border between efficiently solvable and efficiently unsolvable. Theoreticians have this new complexity theory ready in their drawers.

The ability to implement efficient factorization would essentially change cryptography. The current public-key cryptosystems would not be secure for an adversary who owns a quantum computer for factorizing large numbers. Is this really the only possible outcome of building larger quantum computers? Maybe we could build quantum computers working with a few hundred bits, but not with a few thousand bits. Then, we could continue to successfully use public-key cryptosystems by working with numbers consisting of a few thousand digits. It does not matter what the future brings, we do not interpret new discoveries and the resulting progress as a negative development. One has react it positively even if the current public-key cryptosystems will become insecure. Each discovery also opens new possibilities. Quantum effects enable us to build cryptosystems that satisfy very high security requirements. The idea is based on the possibility of bringing two particles into the so-called EPR superposition. This superposition has the special property that, independently of the distance between these two particles, if one measures the state of the first particle, then any later measurement of the second particle gives ultimatively the same outcome as the outcome of the measurement of the first particle. In this way, two parties can agree on the same random bit and use it as a key. One can view an application as follows. The sender has a particle and the receiver has another particle. These two particles are in the EPR superposition. Both sender and receiver perform a measurement on their particle. It does not matter in which order they measure. The only important fact is that they get the same outcome, because the first measurement already results in the collapse of both particles to the same classical state. Then, the second measurement of the classical state can provide only this classical state. If the receiver and the sender generate a common sequence of random bits in this way, they may use it as a key for a symmetric cryptosystem. This quantum effect was predicted many years ago. Albert Einstein did not believe in the experimental confirmation of this prediction. His argument was that this effect, called teleportation, contradicts the locality laws of physical effects. The experimental goal was to show that after the measurement on the first particle the state of the second particle collapsed into the corresponding classical state before the light from the location of the first particle reached the location of the second particle. In this way, one wants to experimentally express the impossibility of influencing the state of one particle by "sending" the information about the change of the state of another particle, because the speed of light is considered to be the upper bound on the speed of everything. To perform such an experiment is very hard, due to the extremely high speed of light the time measurements must be very fine. In spite of the difficulty of time measurement, experimental physicists were able to perform it. This experiment was successfully performed over a distance of 600m across the Danube near Vienna.

The time of pioneers is not over. Science often raises our pulse rates and we are looking forward to spectacular discoveries and pure miracles.

Solutions to Some Exercises

Exercise 9.1 One considers the following superposition:

$$\frac{1}{2}\sqrt{3} \cdot |0\rangle - \frac{1}{2} \cdot |1\rangle \; .$$

Clearly,

$$\left(\frac{1}{2}\sqrt{3}\right)^2 = \frac{1}{4} \cdot 3 = \frac{3}{4} \; , \text{ and } \left(-\frac{1}{2}\right)^2 = \frac{1}{4} \; .$$

Hence, the classical value $|1\rangle$ is measured with the probability $\frac{1}{4}$. We see that the superposition

$$\left(-\frac{\sqrt{3}}{2}\right) \cdot |0\rangle + \frac{1}{2} \cdot |1\rangle$$

satisfies our requirements. One can think about what other superpositions $\alpha \cdot |0\rangle + \beta \cdot |1\rangle$ also satisfy the constraints $\alpha^2 = 3/4$ and $\beta^2 = 1/4$. We see that we are unable to recognize (by measurement) the superposition of the measured quantum system.

Exercise 9.2 For three bits, one has $2^3 = 8$ possible different contents

$$000, 001, 010, 011, 100, 101, 110, \text{ and } 111 .$$

Hence, each state of a quantum register with 3 bits is a superposition

$$\alpha_0 \cdot |000\rangle + \alpha_1 \cdot |001\rangle + \alpha_2 \cdot |010\rangle + \alpha_3 \cdot |011\rangle$$
$$+\alpha_4 \cdot |100\rangle + \alpha_5 \cdot |101\rangle + \alpha_6 \cdot |110\rangle + \alpha_7 \cdot |111\rangle$$

of 8 classical states, where

$$\alpha_i^2 \leq 1 \text{ for } i = 0, 1, \ldots, 7 \text{ and } \sum_{i=0}^{7} \alpha_i^2 = 1 .$$

a) If all 8 classical states have the same probability of being measured, then $\alpha_0^2 = \alpha_1^2 = \alpha_2^2 = \ldots = \alpha_7^2 = \frac{1}{8}$ must hold. Since

$$\frac{1}{8} = \frac{2}{16} = \left(\sqrt{\frac{2}{16}}\right)^2 = \left(\frac{\sqrt{2}}{4}\right)^2 ,$$

one obtains

$$\alpha_0 = \alpha_1 = \alpha_2 = \ldots = \alpha_7 = \frac{\sqrt{2}}{4} ,$$

as a possibility. Can you propose another possibility?

b) Besides the contents $|111\rangle$ and $|000\rangle$ of three bits, there are still six further possible contents. If $|000\rangle$ is observed with probability $\frac{1}{2}$ and $|111\rangle$ is observed with probability $\frac{1}{4}$, then the common probability of the remaining six events is $\frac{1}{4}$ (the sum of all probabilities must be 1). The probability $\frac{1}{4}$ has to be equally distributed on all remaining six contents. Hence, the probability of measuring each of the remaining contents has to be

$$\frac{1/4}{6} = \frac{1}{24} = \frac{6}{144} .$$

The following superposition fulfills our requirements on the probabilities of measuring particular classical states:

$$\frac{1}{2} \cdot |000\rangle + \frac{1}{12}\sqrt{6} \cdot |001\rangle + \frac{1}{12}\sqrt{6} \cdot |010\rangle + \frac{1}{12}\sqrt{6} \cdot |011\rangle +$$
$$\frac{1}{12}\sqrt{6} \cdot |100\rangle + \frac{1}{12}\sqrt{6} \cdot |101\rangle + \frac{1}{12}\sqrt{6} \cdot |110\rangle + \frac{1}{2}\sqrt{2} \cdot |111\rangle .$$

Exercise 9.4 One multiplies

$$H_2 \cdot \begin{pmatrix} \alpha \\ \beta \end{pmatrix} = \begin{pmatrix} \frac{1}{\sqrt{2}} & \frac{1}{\sqrt{2}} \\ \frac{1}{\sqrt{2}} & -\frac{1}{\sqrt{2}} \end{pmatrix} \cdot \begin{pmatrix} \alpha \\ \beta \end{pmatrix}$$
$$= \begin{pmatrix} \frac{1}{\sqrt{2}} \cdot \alpha + \frac{1}{\sqrt{2}} \cdot \beta \\ \frac{1}{\sqrt{2}} \cdot \alpha - \frac{1}{\sqrt{2}} \cdot \beta \end{pmatrix} = \begin{pmatrix} \gamma \\ \delta \end{pmatrix}$$

Our goal is to show that

$$\gamma \cdot |0\rangle + \delta \cdot |1\rangle$$

is a superposition, i.e., that $\gamma^2 + \delta^2 = 1$ holds.

$$\begin{aligned}
\gamma^2 + \delta^2 &= \left(\frac{1}{\sqrt{2}} \cdot \alpha + \frac{1}{\sqrt{2}} \cdot \beta\right)^2 + \left(\frac{1}{\sqrt{2}} \cdot \alpha - \frac{1}{\sqrt{2}} \cdot \beta\right)^2 \\
&= \frac{\alpha^2}{2} + 2 \cdot \frac{1}{\sqrt{2}} \cdot \frac{1}{\sqrt{2}} \cdot \alpha \cdot \beta + \frac{\beta^2}{2} + \frac{\alpha^2}{2} - 2 \cdot \frac{1}{\sqrt{2}} \cdot \frac{1}{\sqrt{2}} \cdot \alpha \cdot \beta + \frac{\beta^2}{2} \\
&= \alpha^2 + \beta^2 \, .
\end{aligned}$$

Since we assumed that $\alpha \cdot |0\rangle + \beta|1\rangle$ is a superposition, $\alpha^2 + \beta^2 = 1$ holds. Therefore, $\gamma^2 + \delta^2 = 1$ and the vector $(\gamma, \delta)^{tr}$ represents a superposition of $|0\rangle$ and $|1\rangle$.

Life can be understood only seen from backward,
but it can be lived only by looking forward.

Søren Kierkegaard

Chapter 10

How to Make Good Decisions for an Unknown Future or How to Foil an Adversary

10.1 What Do We Want to Discover Here?

The computing tasks considered until now were of the following kind: Given a problem instance (or a concrete question), one has to compute a solution (or a correct answer). This means that one has from the beginning the full information (the whole input) that one needs to compute a solution. There are many applications where, at the beginning, one has only partial information about a problem instance before the next part of the input is available.

We illustrate this using the following example. Consider a service center, for instance a hospital with doctors on call. Each doctor

J. Hromkovič, *Algorithmic Adventures*, DOI 10.1007/978-3-540-85986-4_10,

has an emergency doctor's car. When called, the center is required to send a doctor to the accident or to the home of the patient. The center can try to coordinate the movement of the emergency doctors' cars in such a way that some parameters are optimized. For instance, one can try to

- minimize the average waiting time for a doctor,
- minimize the longest possible waiting time,
- minimize the overall length of all trips driven.

To achieve its goal, the center is free to make some decisions. If a new call arrives, it can send a new doctor from the hospital to the accident or send a doctor finishing her or his job at another accident. If the doctor has finished the work at an incident, the center may ask her or him to come back or wait where she or he is for another task, or even take a completely new strategic waiting position. For classical optimization problems, one knows from the beginning all accident locations, accident times, and time intervals sufficient and necessary to treat real patients, and the task is to make a scheduling for the doctors in such a way that the given parameters are optimized. Clearly, one cannot have, in reality, such knowledge about the future. The center does not have a premonition of the place and the time of the next accident. In spite of this uncertainty, which makes finding an optimal solution impossible, one requires from the center a reasonable decision strategy that provides solutions that are good with respect to the knowledge available. Such problems are called **online problems** and the corresponding solution strategies are called **online algorithms**.

One can imagine that there are many problems such as the one presented above. For instance, controlling a taxi center or a police station is very similar to the previous example. In industry, one has to assign workers and machines to execute jobs without knowing what kinds of contracts will be obtained in the near future, and how large, how lucrative, and how urgent they will be. Such optimization problems are called **scheduling**. This kind of online problem is usually very hard, because the future can be very mean.

One calls a doctor back to the hospital, and when the doctor arrives there an emergency call comes from a location neighboring the location of the previous accident. Finding a good optimization strategy under these circumstances can be very hard. Often there does not exist any strategy that would guarantee reasonable solutions independently of future developments. Sometimes, at first glance, one would mean that there is no chance to play against an unknown future, and yet there exist online algorithms that are able to find nearly optimal solutions for each future scenario. Such algorithms are the miracles of this part of algorithmics. The aim of this chapter is to show you such a miracle.

The next section starts with modelling of online problems and introduces the measurement of quality of online algorithms. Here we present an online problem as an example, one where one cannot foil the unknown future.

In Section 10.3, we present an online scheduling problem, for which each deterministic online strategy risks to take a decision that can be very far from optimal. Then, our trump card, we design a randomized online algorithm that ensures a nearly optimal solution for each future development with high probability. As usual, we finish this chapter by giving a summary and solutions to some exercises.

10.2 Quality Measurement of Online Algorithms and a Game Against a Mean Adversary

The problems considered here are so-called **optimization problems**. For each problem instance I, there are potentially many solutions that are called **feasible solutions for I**. Remember the travelling salesman problem. For each complete network of n cities with a direct connection between each pair of cities, there are $(n-1)!/2$ Hamiltonian tours[1]. Each Hamiltonian tour corresponds

[1] Remember that a Hamiltonian tour is a tour starting at some point, visiting each city exactly once, and then returning to the starting point.

to a feasible solution, and so we have a huge number of them. The task is not only to find a feasible solution but a solution whose cost is minimal or at least not far from the cost of an optimal solution. All online problems are optimization problems and the task is to satisfy all requirements provided step by step and finally calculate a feasible solution for the whole input (whole set of requirements). In our example, the emergency center of a hospital is required to handle all emergencies (assuming the capacity of the hospital is sufficient for that). If one is able to do that, then one produces a feasible solution that is described by a sequence of instructions to the doctors. It does not matter which parameters (waiting times, costs, etc.) one optimizes, if one knows the future with all accident locations and accident times one can theoretically[2] always compute an[3] optimal solution. Without knowing the future (i.e., the requirements formulated later), it may happen that we take decisions and put them into practice, which makes them irreversible. The consequences of these decisions may hinder us from satisfying later requirements in an efficient way. The fundamental question posed is the following one:

> *How good can an online algorithm (that does not know the future) be in comparison with an algorithm that knows the whole problem instance (future) from the beginning?*

The answer may vary depending on the problem considered. To be able to study the quality of online algorithms, one has first to understand the meaning of

> *"to be good relative to an algorithm that knows the future."*

To measure the quality of an online algorithm, we use a concept similar to the idea of approximation algorithms. Let I be an instance of a considered optimization problem U. Assume that U is a minimization problem. Let

$$\mathbf{Opt}_U(\boldsymbol{I})$$

---

[2] We write "theoretically" because we do not concern ourselves with the computational complexity. In practice it may happen that we cannot do it because of the limited computing resources.

[3] We write "an" because there may exist several optimal solutions.

denote the cost of an optimal solution for I. Let A be an online algorithm for U that, for each problem instance I of U, computes a feasible solution

$$\mathbf{Sol}_{A}(\boldsymbol{I})\ .$$

The cost of this solution is denoted by

$$\mathbf{cost}(\mathbf{Sol}_{A}(\boldsymbol{I}))\ .$$

We define the **competitive ratio** $\mathbf{comp}_{A}(\boldsymbol{I})$ **of** $\boldsymbol{A}$ on the instance $\boldsymbol{I}$ as

$$\mathbf{comp}_{A}(\boldsymbol{I}) = \frac{\mathbf{cost}(\mathbf{Sol}_{A}(\boldsymbol{I}))}{\mathbf{Opt}_{U}(\boldsymbol{I})}.$$

In this way, $\text{comp}_A(I)$ says by how many times the cost of the solution $\text{Sol}_A(I)$ of A on I is larger (worse) than the best possible cost. For instance, if $\text{Opt}_U(I) = 100$ (i.e., the optimal solutions for I have cost 100), and $\text{cost}(\text{Sol}_A(I)) = 130$ (i.e., the online algorithm A computes a solution with the cost 130), then

$$\text{comp}_A(I) = \frac{130}{100} = 1.3$$

meaning that the computed solution is 1.3 times worse than the optimal solution. One can also say that $\text{Sol}_A(I)$ costs 30% more than the optimal solutions.

Now we know how to determine how good an online algorithm is for a problem instance I. We measure the quality of A with respect to a guarantee that is provided by A in any[4] case. Therefore, we define the quality of A as follows:

We say that A is a **δ-competitive online algorithm** for U, if, for all problem instances I of U,

$$\text{comp}_A(I) \leq \delta\ .$$

If $\text{comp}_A(I) \leq 1.3$ for all problem instances I of U (i.e., if A is 1.3-competitive), then it means that, for any problem instance, we

[4] In computer science, one speaks about the "worst case".

compute a solution whose cost is not larger than 130% of the cost of an optimal solution. For many input instances it may happen that our solutions are even essentially closer to the optimal ones. For several hard problems, one should be reasonably happy with such a guarantee.

Exercise 10.1 Let $\mathrm{Opt}_U(I) = 90$ for a problem instance I. Assume our online algorithm computes a solution $\mathrm{Sol}_A(I)$ with $\mathrm{cost}(\mathrm{Sol}_A(I)) = 135$.

a) Calculate the competitive ratio of A on I.
b) How many percent higher cost than the optimal one has the computed solution $\mathrm{Sol}_A(I)$?

Exercise 10.2 For any problem instance I, we introduced the values $\mathrm{Opt}_U(I)$ and $\mathrm{cost}(\mathrm{Sol}_A(I))$. What does the following number express?

$$\frac{\mathrm{cost}(\mathrm{Sol}_A(I)) - \mathrm{Opt}_U(I)}{\mathrm{Opt}_U(I)} \cdot 100$$

Assume one designs an online strategy A for an optimization problem U and aims to estimate $\mathrm{comp}_A(I)$. Computer scientists say that they want to **analyze** the competitive ratio of A. This may require a lot of hard work. We know cases of problems and designed online algorithms for them where, after an effort of several years, one is unable to approximate comp_A. The hardness of this analysis is related to the fact that one has to estimate the maximal value of $\mathrm{comp}_A(I)$ over all infinitely many problem instances I.

To analyze comp_A, the researchers use a helpful game between an algorithm designer and his adversary. The aim of the algorithm designer is to design an online algorithm. The adversary tries to prove that the designed algorithm is not good enough by designing problem instances for which the algorithm does not work well (Fig. 10.1).

In this game, one can view the algorithm designer as an enthusiastic optimist who is happy about the product of her or his work. The adversary can be viewed as a confirmed skeptic who questions all products of the algorithm designer and tries to convince everybody about the weakness of the designed algorithm. Good research teams need both, optimists and pessimists. In this way, one can

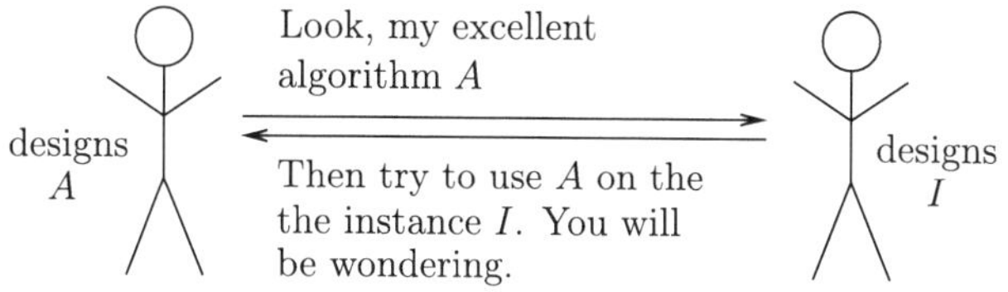

designs A

Look, my excellent algorithm A

Then try to use A on the the instance I. You will be wondering.

designs I

Fig. 10.1

produce original ideas with considerable enthusiasm, check them carefully, and finally improve them.

In the analysis of online algorithms, one views the adversary as a **mean** person. This is related to the actual situation. The adversary knows the algorithm A and is allowed to design the future in such a way that A is not successful. Since the adversary knows A, he knows exactly which partial solutions are produced by A for a given part (prefix) of the input. Therefore, the game can be viewed as follows. The adversary shows a part of the future and waits for the action of A. After that the adversary estimates the next part of the future. After examining how A acts on that, he builds the next requirements. Hence, the adversary has a good possibility of leading A astray. If the adversary succeeds, he has proved that comp_A cannot be good.

We use the following example to hit three flies at once. First, we illustrate the definition of the competitive ratio using a concrete scenario. Second, we transparently show how a mean adversary proves the limits of any possible online algorithm for the problem considered. Finally, we learn that there are computing tasks that cannot be solved successfully using any online algorithm.

Example 10.1 Paging

Paging is a well-known problem that has to be incessantly solved in all computers. Each computer has two kinds of memory. One is small and the other is large. The small memory is called **cache** and the computer has random access to it, which means that the access to the data is extremely fast. The large memory is called the **main memory** and it is considerably larger than the cache (Fig. 10.2).

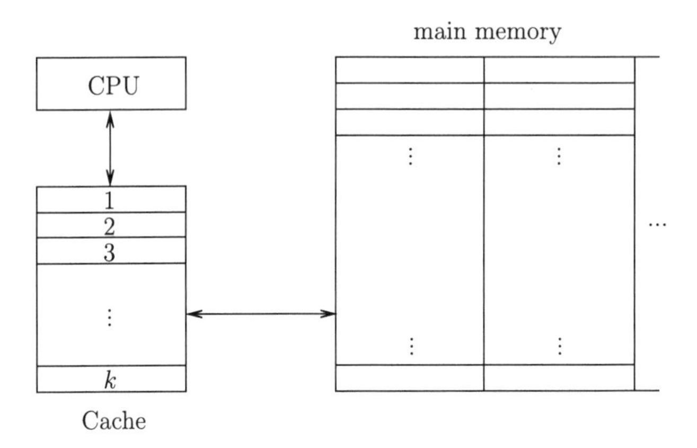

Fig. 10.2

The access to the data in the main memory is slow. In fact the computer directly proceeds on data in cache only. If one wants to work with or see data that is not in cache, one has to transfer this data to the cache first and then one can read them. This data transfer is time consuming. Therefore, one strives to have all data needed in the near future in cache. But one cannot ensure this, because the computer does not know the future, i.e., it does not know which data the user wants to access next. The computer only receives from time to time a requirement to show some data or to proceed on some data. If these data are not in cache, it must move them there. Since the cache is usually full, at first, the computer has to delete or transfer some part of the data in cache to the main memory. Now one understands the framework for an online strategy. Which data have to be removed from cache? Please, remove any except those needed in the next application.

Let us model this situation more precisely. Both memories are partitioned into data blocks called **pages**. Cache can contain at most k pages for a fixed number k. Usually, cache contains exactly k pages, i.e., it is full. The main memory contains all data needed and so the amount of data there may be very large. One can transfer only whole pages between cache and the main memory. This is similar to a thick book, in which one can see (open) at most k pages at any time. If one wants to see a new page, one has to close one of the open pages. Therefore, one calls this task paging.

A complete instance I of paging can be, for instance,

$$I = 3, 107, 30, 1201, 73, 107, 30 .$$

This instance corresponds to the requirement to see the pages with numbers $3, 107, 30, 1201, 73, 107, 30$ one after each other in the prescribed order. The pages 107 and 30 even have to be visited twice. Page 107 has to be open in the second time period and in the sixth time period. This means it would be wrong to remove page 107 from cache after reading it in the second step, because one needs it in the sixth time period.

Because one considers paging as a minimization problem, one has to assign a cost to each feasible solution. Since the time necessary for transporting a page from the main memory to cache is incomparably higher than the time sufficient for accessing data in cache, one is allowed to measure the cost as follows:

- the cost for a direct access to cache is set to 0, and
- the cost for transferring a page from the main memory to cache is set to 1.

Consider the following situation. One has a cache of size 3 (i.e., $k = 3$) that contains three pages $5, 30$, and 107. The problem instance given is $I = 3, 107, 30, 1201, 73, 107, 30$. The following optimal solution has cost 3. First, the computer sends page 3 to cache and sends page 5 from cache to the main memory. We denote this exchange of pages 3 and 5 between cache and the main memory by

$$5 \leftrightarrow 3 .$$

The next pages asked for are pages 107 and 30, which are already available in cache. After reading them, the computer moves page 1201 to cache from the main memory by exchanging it with page 3. Pages 30 and 107 remain in cache because the computer knows that it will need them soon. In the fifth step, the computer exchanges page 1201 from cache for page 73. The last two requirements, 107 and 30, can be handled by direct access in cache. The above described plan (solution) can be represented as follows:

$$5 \leftrightarrow 3, \bullet, \bullet, 3 \leftrightarrow 1201, 1201 \leftrightarrow 73, \bullet, \bullet$$

The symbol $\bullet$ stands for no action between cache and the main memory, and the denotation $a \leftrightarrow b$ represents the exchange of page a in cache for page b from the main memory. Table 10.1 shows the content of cache in the 8 time periods considered.

time	0	1	2	3	4	5	6	7
operation		$5 \leftrightarrow 3$	$\bullet$	$\bullet$	$3 \leftrightarrow 1201$	$1201 \leftrightarrow 73$	$\bullet$	$\bullet$
	5	3	3	3	1201	73	73	73
cache	30	30	30	30	30	30	30	30
	107	107	107	107	107	107	107	107
read		3	107	30	1201	73	107	30

Table 10.1

One is allowed to claim that the above presented solution is optimal. At the beginning the required pages 3, 1201, and 73 are not in cache. Following the requirements given by the instance $I = 3, 107, 30, 1201, 73, 107, 30$ it is obvious that these three pages have to be transferred to cache and so any solution must have a cost of at least 3.

Exercise 10.3 Find optimal solutions for the following instances of paging:

(a) $k = 3$, cache contains $1, 2, 3$ and $I = 7, 9, 3, 2, 14, 8, 7$
(b) $k = 5$, cache contains $1, 101, 1001, 1002, 9$ and $I = 1002, 7, 5, 1001, 101, 3, 8, 1, 1002$

Note that if one is forced to bring a new page to cache, one has k possibilities for choosing the page that has to be exchanged for the new one. Hence, after a few steps, the number of different possibilities grows very quickly and so searching for an optimal solution may become confusing. Nevertheless, an optimal solution can always be calculated, and everything is only a question of the amount of work. Considering paging as an online problem, the situation changes dramatically. The requirements to read a page come separately one after another. Only after fulfilling a requirement through a possible exchange of pages between cache and the main memory, will the next requirement become public. The adversary can be really mean. He forces reading of exactly the page that was moved to the main memory in the last step. It does not matter what online strategy for the exchange of pages is used, the adversary always forces us to read exactly the last removed page. In this way, for each particular requirement of the instance constructed by the adversary, a communication between cache and the main memory takes place. Hence, the cost of the solution is the length of the instance I, and so the cost is the maximum possible.

Let us play the game with the adversary for a concrete example. Let $k = 3$ and cache contains pages $1, 2$, and 3. At the beginning, the adversary forces us to read page 4. To bring page 4 to cache, one has to remove the pages in cache. Assume page 2 is removed. Then, in the next step, the adversary asks for page 2. Page 2 is not in cache anymore and must be transferred to cache. Assume our online strategy acts through $4 \leftrightarrow 2$. Then, the adversary asks for page 4 in the next step. If the online strategy had decided to perform $1 \leftrightarrow 4$, and did it, the adversary would ask for page 1. In this way, the adversary creates the instance

$$4, 2, 4, 1,$$

and forces the online strategy to determine the feasible solution

$$2 \leftrightarrow 4, 4 \leftrightarrow 2, 1 \leftrightarrow 4, 4 \leftrightarrow 1 \ .$$

This solution has the maximal possible cost 4. The optimal solution for the instance $4, 2, 4, 1$ is

$$3 \leftrightarrow 4, \bullet, \bullet, \bullet$$

of cost 1. But one can calculate it only if the instance $4, 2, 4, 1$ of paging is known from the beginning.

Exercise 10.4 Consider the following online strategy. One always removes the page with the smallest number. Consider $k = 4$ and let cache contain pages $1, 3, 5$, and 7 at the beginning. Play the role of the adversary and design a problem instance of 10 requirements (i.e., of length 10), such that the solution of the online strategy has the maximal possible cost 10.

Exercise 10.5 Consider the online algorithms that always remove one of those pages from cache that were read (asked for) the fewest number of times until now. If several such pages are in cache, the online strategy removes the one with the largest number. Play the adversary for this strategy and use the following starting situations and goals:

(a) $k = 4$, cache contains pages $1, 2, 3$, and 4, the instance I constructed has to be of length 4, and $\mathrm{Opt}_{paging}(I) = 1$.
(b) $k = 5$, cache contains $1, 7, 103, 5, 9$, and $\mathrm{Opt}_{paging}(I) = 2$ for the designed instance I of paging.

We generalize our experience and show that each online algorithm for cache of k pages fulfills

$$\mathrm{comp}_A \geq k \ .$$

Hence, there is no good online strategy that can work satisfactorily for all instances of paging.

Assume without loss of generality that pages $1, 2, 3, 4, \ldots, k$ are in cache of size k. Let A be an arbitrary online algorithm for paging. The adversary starts to construct an instance by asking for page $k + 1$. Since page $k + 1$ is not in cache, A has to exchange one of its pages for page $k + 1$. Assume A performs the action

$$s_1 \leftrightarrow k + 1$$

where s_1 is from $\{1, 2, \ldots, k\}$. Now cache contains pages

$$1, 2, \ldots, s_1 - 1, s_1 + 1, \ldots, k, k + 1 \ .$$

The designer continues to build the hard instance of paging for A by asking for page s_1. Since page s_1 is not in cache A is forced to exchange one of the pages in cache for s_1. Assume A performs

$$s_2 \leftrightarrow s_1$$

for a page s_2 from $\{1, 2, \ldots, k, k+1\} - \{s_1\}$. After this exchange, cache contains pages $1, 2, \ldots, s_2 - 1, s_2 + 1, \ldots, k+1$, i.e., all pages with a number i from $\{1, 2, \ldots, k+1\} - \{s_2\}$. As expected, the adversary asks for page s_2, which was removed from cache in the last step. In this way the adversary can proceed until an instance

$$I_A = k+1, s_1, s_2, \ldots, s_{k-1}$$

of length k is created and

$$\text{cost}(\text{Sol}_A(I_A)) = k$$

holds. Observe that, for different strategies, different instances can be constructed. Now we claim that

$$\text{Opt}_{paging}(I_A) = 1 \ .$$

Let us argue for that. If one knows the whole instance I_A from the beginning, one can proceed as follows. The numbers $s_1, s_2, \ldots, s_{k-1}$ are all from $\{1, 2, \ldots, k+1\}$ and they are $k-1$ in number. Hence, there is a number j in $\{1, 2, \ldots, k\}$ that is not among the numbers $s_1, s_2, \ldots, s_{k-1}$. If one takes the action

$$j \leftrightarrow k+1$$

at the beginning when page $k+1$ is required, then one does not need any communication between cache and the main memory later, since all pages $s_1, s_2, \ldots, s_{k-1}$ asked for in the next $k-1$ steps are in cache. One never misses page j. In this way

$$j \leftrightarrow k+1, \bullet, \bullet, \ldots, \bullet$$

is the optimal solution for I_A. For instance, if $k = 4$ and $I_A = 5, 3, 1, 4$, then $j = 2$ and

$$2 \leftrightarrow 5, \bullet, \bullet, \bullet$$

is the optimal solution. Since $\text{Opt}_{paging}(I_A) = 1$, one obtains

$$\text{comp}_A(I_A) = \frac{\text{cost}(\text{Sol}_A(I_A))}{\text{Opt}_{paging}(I_A)} = \frac{k}{1} = k$$

for each online algorithm A. Hence, there does not exist any δ-competitive online algorithm for paging with cache of size k and $\delta < k$.

What to do with paging in practice? The answer is too complicated to be explained here in detail. The idea is to design online algorithms that behave well for at least the typical (frequently occurring) instances and so achieve a good competitive ratio on average. Using extensive experiments, one recognized that the probability of asking for pages read recently is higher than asking for pages used only a few times up to now. One can use this fact and randomization in order to design practical online algorithms for paging. □

10.3 A Randomized Online Strategy

In this section, we want to show that one can find good online strategies in cases that seem hopeless at first glance. There exist online computing tasks for which one can unexpectedly take decisions that are almost as good as those made by somebody who knows the future.

To remain transparent, so without using too much mathematics, we consider a very simple version of scheduling. Assume one has a factory with n different work stations. Each work station consist of a machine or a collection of machines that can perform specific tasks. The factory receives tasks from its customers. Each task specifies which stations are required and in which order for performing the required processing. In this simplified version one assumes that each task requires all n stations and the processing takes the same time unit on each of the stations. The only free parameter of customer choice is the order in which the stations have to be used. For instance, a factory can have $n = 5$ stations S_1, S_2, S_3, S_4, and S_5. A task

$$A = (1, 3, 5, 4, 2)$$

means that the customer needs to use the stations in the order

$$S_1, S_3, S_5, S_4 \text{ and } S_2 \ .$$

The aim of the factory is to perform each task as soon as possible. If the task A is the only one the factory receives, the factory can perform it in the minimal time of 5 time units[5]. In the first time unit S_1 is active, in the second time unit station S_3 works, in the third time unit S_5 is used, in the fourth time unit S_4 is applied, and finally in the fifth time unit station S_2 is used. The problem is that several customers can be concurrent. If several customers ask for the same station in the same time unit, only one can be satisfied and the others have to wait. Here we investigate the simplest version of the problem with only two customers. The goal of the company is to minimize the overall time for the complete processing of both tasks given by the customers. For instance, consider $n = 4$ and the following two tasks:

$$A_1 = (1, 2, 3, 4)$$
$$A_2 = (3, 2, 1, 4) \ .$$

The factory can proceed as follows. In the first time unit, both tasks can be performed **in parallel**. Simply station S_1 performs the first job of the first task A_1, and station S_3 performs the first job of A_2. After the first time unit, both first jobs of A_1 and A_2 are finished and both tasks require station S_2 for their second job. But this cannot be done simultaneously (in parallel), since each station can perform at most one job in a time unit. Hence, the factory has to decide which job will be performed and which task has to wait without any progress in its processing. Assume the company decides to assign S_2 to the second job of A_1 and lets A_2 wait. Since the jobs of A_2 must be processed in the prescribed order, there is no work on A_2 in this time period and we say that A_2 got a **delay** of size one time unit. After the second time unit, the first two jobs of A_1 and the first job of A_2 are ready (Table 10.2). Now, A_1 requires S_3 and A_2 requires S_2 again. Now the factory can satisfy both wishes at

[5] one unit for each station

once, and so it decides to proceed with A_1 on station S_3 and with A_2 on station S_2 in parallel. After the third time unit, A_1 asks for S_4 and A_2 asks for S_1. The factory can satisfy them in parallel. Hence, after the fourth time unit, the work on A_1 is finished and it remains to finish the work on A_2 by using station S_4 in the fifth time unit. Hence, after 5 time units, the work of A_1 and A_2 is over. The solution used is an optimal one, because there is no solution for processing A_1 and A_2 in four time units.

There does not exist any solution of time less than 5 time units, because each of A_1 and A_2 requires 4 time units to be performed and this would be possible only if in each time unit both tasks were processed in parallel. Hence, at the beginning A_1 must be processed on S_1 and A_2 on S_3, and so the **collision** of the same requirement for S_2 in the second time unit cannot be avoided. Therefore, one of A_1 and A_2 is forced to wait in the second time unit and so cannot be finished faster than in 5 time units.

Table 10.2 transparently presents the processing of tasks A_1 and A_2 as described above. Each column corresponds to a time unit and shows which stations are active in this time unit. Viewing the rows, one can observe the progress in processing A_1 and A_2 step by step.

time units	1	2	3	4	5
A_1	S_1	S_2	S_3	S_4	
A_2	S_3		S_2	S_1	S_4

Table 10.2

Exercise 10.6 Assume that in the first time unit the company decides to work on A_1 only and lets A_2 wait in spite of the fact that the required station S_3 for A_2 is free. Is it possible to finish both tasks in altogether 5 time units? Describe your solution for scheduling of the stations in a way similar to Table 10.2.

Exercise 10.7 Assume the factory has $n = 6$ solutions and has to work on two tasks $A_1 = (1, 2, 3, 4, 5, 6)$ and $A_2 = (1, 3, 2, 6, 5, 4)$. How many time units are sufficient and necessary to perform both tasks? Depict your solution in a way similar to Table 10.2.

There is a graphic representation of this problem that makes searching for a solution transparent. For a factory with n stations, one draws an $n \times n$-field. The jobs of task A_1 are assigned to the columns in the given order, and the jobs of task A_2 are assigned to the rows. Figure 10.3 shows the 4×4-field for the tasks

$$A_1 = (1, 2, 3, 4) \text{ and } A_2 = (3, 2, 1, 4) .$$

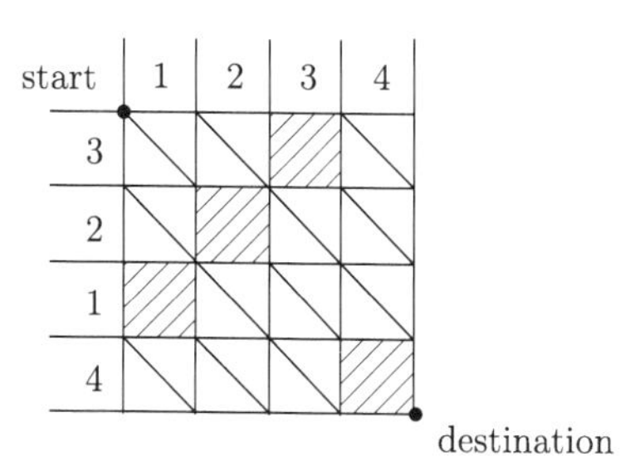

Fig. 10.3

The small fields (squares) in which the two same requirements meet are called **obstacles** and are hatched in Fig. 10.3. For instance, an obstacle is the intersection of the first row and the third column because A_2 asks for station S_3 at the beginning and A_1 asks that S_3 processes its third job. Analogously, an obstacle is in the intersection of the second row and the second column because both A_1 and A_2 ask that the same station S_2 processes their second jobs. Note that the number of obstacles is exactly n because, for each station S_i, there is exactly one row and exactly one column labelled by i (asking for S_i). Therefore, we see in Fig. 10.3 exactly 4 obstacles. For all free squares without obstacles (for all squares with different requirements) we draw a line between the upper-left corner of the square and the bottom-right corner of the square, and call this line a **diagonal edge** (Fig. 10.3). The points where the lines of the field (grid) meet are called vertices (Fig. 10.4).

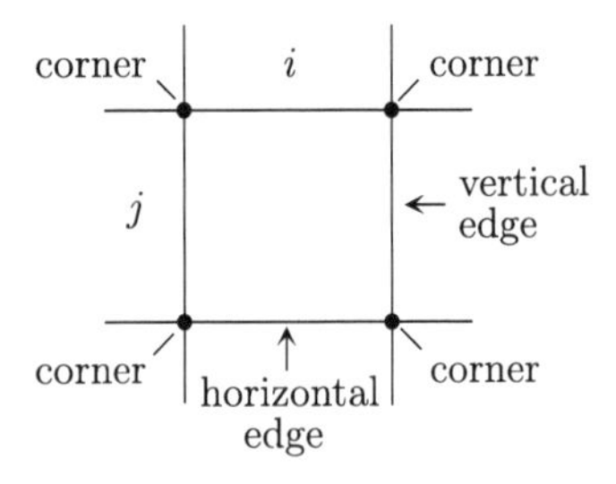

Fig. 10.4

The lines connecting two neighboring vertices of the grid are called **edges** (Fig. 10.4). The edges that are part of the horizontal lines[6] of the field are called **horizontal edges**. Analogously, the edges running in the vertical direction[7] are called vertical edges (Fig. 10.3). Searching for a solution for a problem instance (A_1, A_2) corresponds to searching for a path (route) from the upper-left corner of the whole field (denoted by START in Fig. 10.3) to the bottom-right corner of the field (denoted by DESTINATION in Fig. 10.3). The path goes step by step from a vertex to a vertex.

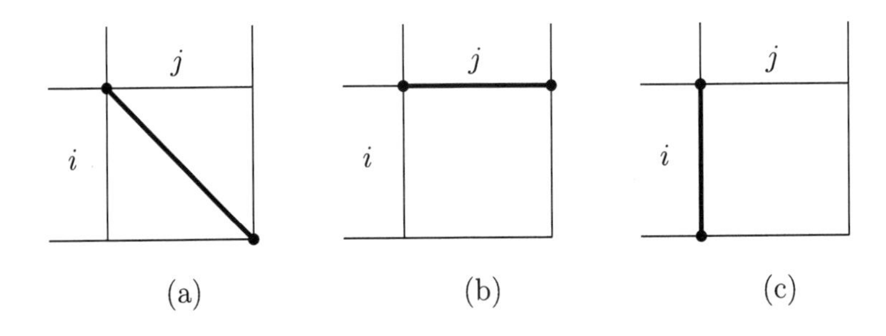

Fig. 10.5

If the path takes a diagonal edge of a square in the intersection of the row labelled by S_i and the column labelled by S_j (Fig. 10.5(a)), then both tasks A_1 and A_2 are processed in parallel on the required stations S_i and S_j. To take a horizontal edge (Fig. 10.5(b)) corresponds to processing A_1 on station S_j. Here, task A_2 is not

[6] that separate two neighboring rows

[7] as part of the vertical lines separating two neighboring columns

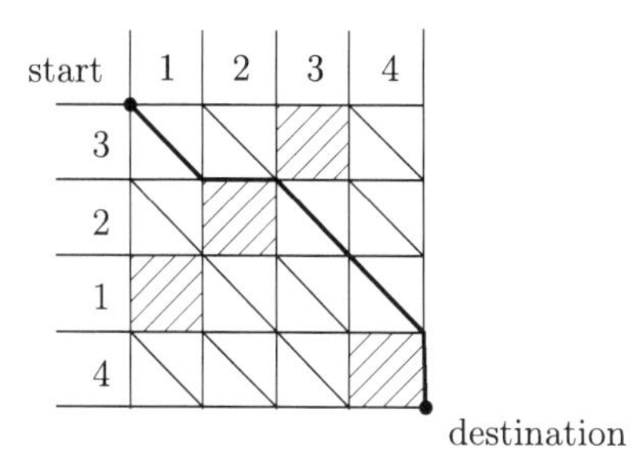

Fig. 10.6

processed and gets delayed in this way. To take a vertical edge (Fig. 10.5(c)) corresponds to processing A_2 on S_i and waiting for A_1. Figure 10.6 depicts the solution described in Table 10.2 for the problem instance (A_1, A_2) from Fig. 10.3. We see that one uses diagonal edges wherever possible. Only when one has to bypass the obstacle in the square $(2, 2)$ does one use the horizontal edge, and at the end when A_1 is ready the vertical edge is used to reach the destination.

The cost of a solution, the number of time units used for processing A_1 and A_2, is exactly the length of the corresponding path from START to DESTINATION. The length of a path is measured as the number of edges of the path. In Fig. 10.6 the path consists of 5 edges and the i-th edge of the path corresponds to assignment of stations in the i-th time unit in Table 10.2. If one knows the complete tasks A_1 and A_2 from the beginning, one can use well-known efficient algorithms for finding the shortest paths between two vertices of a graph in order to calculate the optimal solution for the scheduling of the stations.

Exercise 10.8 Consider the tasks $A_1 = (6, 5, 4, 3, 2, 1)$ and $A_2 = (4, 5, 6, 2, 3, 1)$ for a factory with 6 stations. Depict the corresponding graphical representation of this problem instance with 6 obstacles. Find a shortest[8] path from the start to the destination and use this path to derive a solution representation such as that given in Table 10.2.

We see that this optimization task can be solved easily. But to solve this problem in the online setting is harder. At the beginning, the

[8] there may exist several shortest paths

factory knows only the first jobs of A_1 and of A_2. For instance $A_1 = (3, \ldots)$ and $A_2 = (5, \ldots)$ and the rest is still unknown. If the factory performs the first jobs of A_1 and A_2 on S_3 and S_5 in parallel, the next requirements (jobs) of A_1 and of A_2 are formulated (become known). The rule is that the factory knows for each of the tasks A_1 and A_2 only the next, still unprocessed job, and all further requirements are still unknown.

Now the game between the algorithm designer and his adversary can begin. The algorithm designer suggests an online strategy and the adversary creates a problem instance that is hard for this strategy. The strategy of the adversary is very simple and also general in the sense that it works for any designed online algorithm. The idea is to force any designed online algorithm to use a non-diagonal edge at least in every second step (time unit). This means that in at least every second step, there is a delay in processing one of the two tasks A_1 and A_2. How can the adversary cause that?

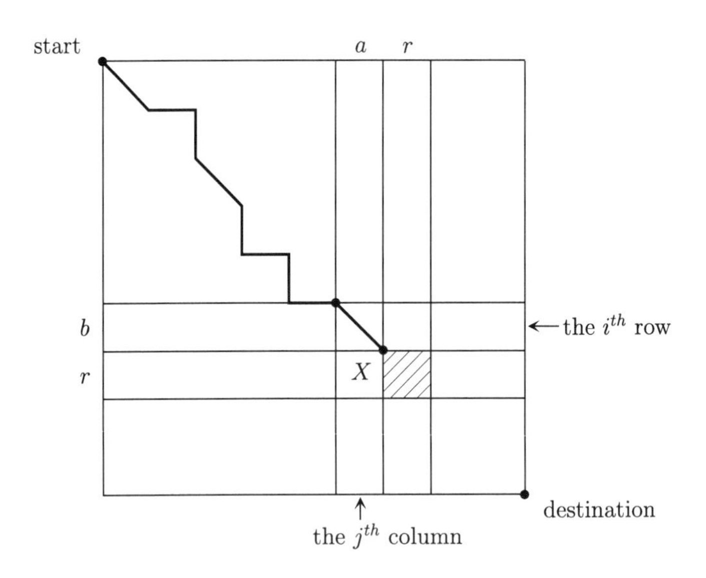

Fig. 10.7

Assume, as depicted in Fig. 10.7, that the last step of the online algorithm used a diagonal edge. If the path in Fig. 10.7 is now in the vertex X, at the intersection of the horizontal line separating the i-th row from the $(i+1)$-th row and the vertical line separating the j-th column from the $(j+1)$-th one, then the first j jobs of A_1 and the first i jobs of A_2 have already been executed. Now, the adversary can choose the $(j+1)$-th requirement (job) of A_1 and the $(i+1)$-th requirement of A_2. The adversary chooses the same requirement S_r for both. He has only to take care to choose S_r in such a way that neither A_1 nor A_2 had this requirement before. In this way, one gets an obstacle (Fig. 10.7) that must be bypassed from the vertex X either in the way depicted in Fig. 10.5(b) or in the way outlined in Fig. 10.5(c).

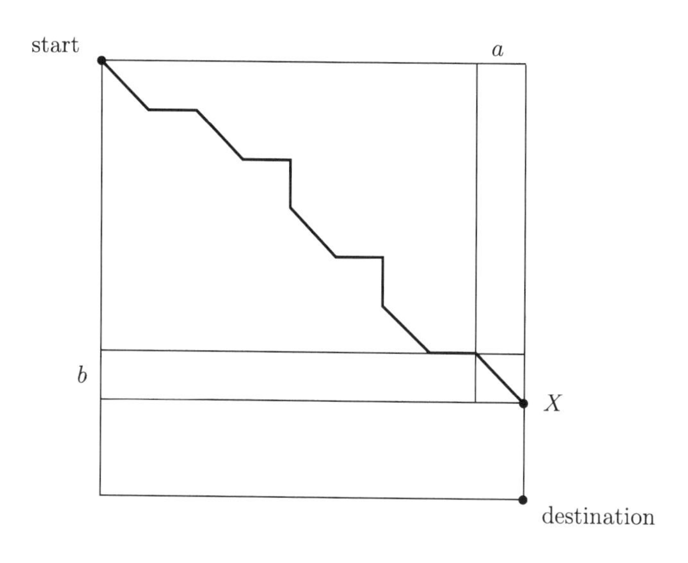

Fig. 10.8

If the online algorithm reaches the border of the field using a diagonal step (Fig. 10.8), then the work of one of the tasks is finished and no diagonal step is possible anymore. Hence, the adversary can order the other non-executed jobs of the remaining task arbitrarily.

What can be concluded? For each online algorithm A, the adversary can construct an input x_A such that A working on x_A uses at least one non-diagonal edge in every second step. This means that A delays the processing of one of the tasks at least in every second step. If one has m stations, then the overall number of delays is at least $m/2$. These $m/2$ delays are somehow distributed among the two tasks A_1 and A_2. Hence, at least one of the tasks had at least $m/4$ delays. We conclude that the overall time for processing x_A using A is at least

$$m + m/4 ,$$

since the shortest possible path from START to DESTINATION consists of m diagonal edges, and one has to add to m the maximum of the number of delays on A_1 or on A_2.

Let us observe the strategy of the adversary for a concrete online algorithm.

Example 10.2 Consider the following online strategy A for the movement in the field from START to DESTINATION.

1. If a path can be prolonged using a diagonal edge, then one takes it.
2. If the path meets an obstacle (i.e., there is no possibility to continue using a diagonal edge), take the vertical or horizontal edge that reaches a vertex that is closer to the main diagonal[9] of the whole field. If both possibilities are equally good, take the horizontal edge.
3. If the path reaches the right border of the field, take the vertical edges to reach DESTINATION.
4. If the path reaches the lower border of the field, take the horizontal edges to reach DESTINATION.

For the online algorithm A, the adversary constructs a hard problem instance $x_A = (A_1, A_2)$ as follows. The adversary starts with $A_1 = 1, \ldots$ and $A_2 = 1, \ldots$. Hence, one has an obstacle at the beginning (Fig. 10.9). Following rule 2, A takes the horizontal edge, because all possibilities of bypassing the obstacles have the same distance to the main diagonal of the field.

[9] the line connecting START and DESTINATION

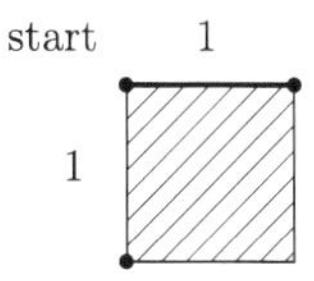

Fig. 10.9

After that, the adversary is asked to pose the second requirement of A_1. Observe that there is exactly one obstacle in each row and in each column. Hence, the adversary cannot lay down an obstacle in this situation, i.e., independently of his choice, the next diagonal edge can be used. Assume the adversary takes S_2 as the second job of A_1. Then, following rule 1, the online algorithm uses the diagonal edge as depicted in Fig. 10.10.

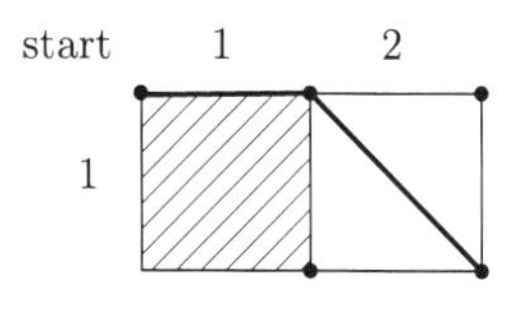

Fig. 10.10

Now, jobs specified up to now are executed, and so the adversary is asked to formulate the next requirements for both A_1 and A_2. This means that the adversary has again the possibility to lay down an obstacle. The adversary continues as follows:

$$A_1 = (1, 2, \mathbf{3}, \ldots) \text{ and } A_2 = (1, \mathbf{3}, \ldots) .$$

Hence, A stands in front of an obstacle (Fig. 10.11) and, following its rule 2, the algorithm A has to take the vertical edge to bypass the obstacle (Fig. 10.11).

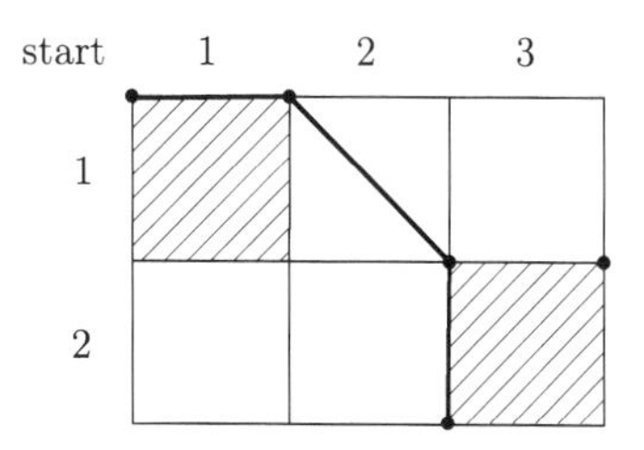

Fig. 10.11

Again, one reaches a situation (Fig. 10.11) where there is no possibility to lay down an obstacle for the path built until now. The adversary specifies $A_2 = (1, 3, \mathbf{2}, \ldots)$. Following rule 1, the online algorithm A takes the diagonal edge (Fig. 10.12), and the path reaches a new row and a new column simultaneously in this way. Hence, the adversary can lay down an obstacle and he does so by using the following specification

$$A_1 = (1, 2, 3, \mathbf{4}, \ldots) \text{ and } A_2 = (1, 3, 2, \mathbf{4}, \ldots) .$$

Following rule 2, the algorithm A bypasses the obstacle by taking the horizontal edge (Fig. 10.12). If the factory has 4 stations, the construction of the whole input $x_a = (A_1, A_2)$ is finished. To reach the destination, one still has to take the vertical edge. In this way, the constructed path contains 6 edges, and only two of them are diagonal edges. One starts with a horizontal edge, and at most every second edge is a diagonal one.

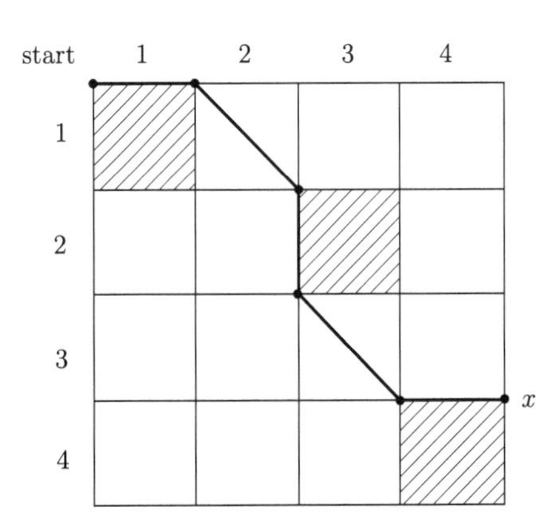

Fig. 10.12

□

Exercise 10.9 Assume the factory has 7 stations. Complete the construction of x_A from Example 10.2. For this case (starting from the situation in Fig. 10.12) by taking on the role of the adversary.

Exercise 10.10 Draw the optimal solution for the problem instance $A_1 = (1, 2, 3, 4)$ and $A_2 = (1, 3, 2, 4)$ in Fig. 10.12.

Exercise 10.11 One changes the online strategy A from Example 10.2 by forcing a bypassing of all obstacles using the corresponding horizontal edge. Play the adversary for the new online strategy A' and construct a hard problem instance for A'.

Exercise 10.12 Consider an online strategy B that, independently of the input, takes 3 vertical edges at the beginning (i.e., let A_1 wait for three time units), and then continues following the strategy of A. Play the adversary for B and construct a hard problem instance for B.

We showed that, for each online algorithm A, one can construct a problem instance x_A such that A needs at least

$$m + m/4$$

time units to process x_A.

Is this good news or bad news? To answer this question, one has to investigate how good solutions can be calculated, if one knows the future (i.e., if one knows the whole input instance from the beginning). We will recognize that each problem instance (A_1, A_2) for m stations can be solved in

$$m + \sqrt{m}$$

time units. Therefore, for each online algorithm A,

$$\text{comp}_A(I) \geq \frac{m + 0.25 \cdot m}{m + \sqrt{m}}$$

holds. For large numbers m it means that the solutions performed in the online manner can be almost 25% more expensive than optimal solutions.

To show that each problem instance can be performed in $m + \sqrt{m}$ time units, we present an argument that is probably unknown for most readers. We consider several algorithms for each instance and recognize that the solutions calculated using these algorithms need $m + \sqrt{m}$ time units on average. If the solutions use $m + \sqrt{m}$ time units on average, then there must be at least one solution among them[10] that needs at most $m + \sqrt{m}$ time units.

[10] If all solutions require more than $m + \sqrt{m}$ for processing the tasks, then the average must be greater than $m + \sqrt{m}$.

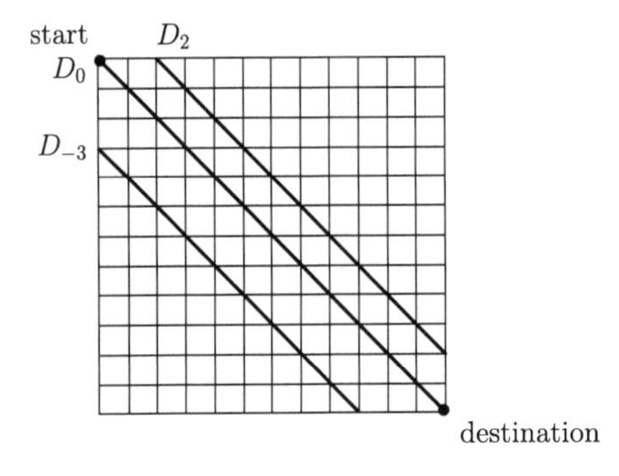

Fig. 10.13

For simplicity, we set $m = k^2$, and so $\sqrt{m} = k$ is an integer. We take $2k + 1$ so-called diagonal strategies into account. By D_0 one denotes the main diagonal of the field (Fig. 10.13) that leads from START to DESTINATION. D_i denotes the diagonal of the whole field that lies i squares (horizontal edges) above D_0. Analogously, D_{-j} denotes the diagonal that lies j squares (vertical edges) below D_0. In Fig. 10.12, one sees the diagonals D_0, D_2, and D_{-3}.

For each diagonal D_l, one considers a strategy SD_l that strives to visit all vertices lying on the diagonal D_l. For each $i \geq 0$, the strategy SD_i first takes i horizontal edges in order to reach the upper-left vertex of the diagonal D_i. After that, SD_i tries to use the diagonal edges of D_i. If this is impossible because of an obstacle, then SD_i takes first the corresponding horizontal edge and then the following vertical edge in order to bypass the obstacle and to reach the next vertex of the diagonal D_i (Fig. 10.14). If SD_i reaches the bottom vertex of D_i, it uses the i vertical edges on the border to reach the destination.

The strategy SD_{-i} starts with taking the i vertical edges on the border in order to reach the upper vertex of D_{-i}. After that SD_{-i} runs analogously to SD_i by trying to use the diagonal edges of D_i. Finally, the i horizontal edges on the border of the field are used to reach the destination.

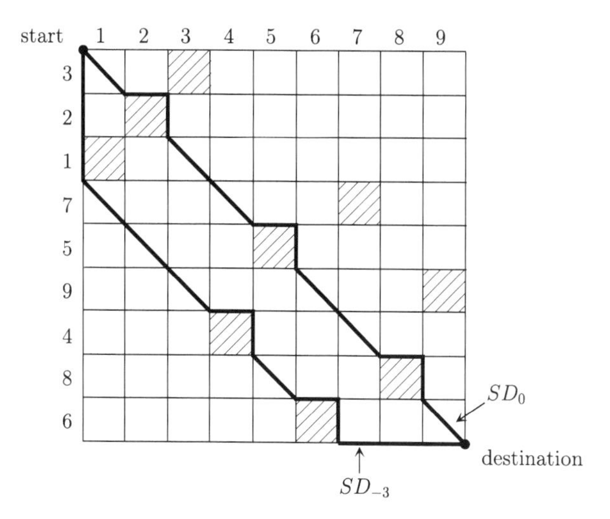

Fig. 10.14

Figure 10.14 shows the solution provided by SD_0 and SD_{-3} for the problem instance

$$A_1 = (1, 2, 3, 4, 5, 6, 7, 8, 9) \text{ and } A_2 = (3, 2, 1, 7, 5, 9, 4, 8, 6) .$$

Exercise 10.13 Estimate the solutions derived by the strategies SD_1, SD_2, and SD_{-2} for the problem instance in Fig. 10.14.

Exercise 10.14 Consider the problem instance $A_1 = (9, 8, 7, 6, 5, 4, 3, 2, 1)$ and $A_2 = (9, 7, 8, 4, 5, 6, 2, 3, 1)$. Draw the solutions calculated by the strategies SD_3, SD_0, SD_{-1}, and SD_{-3}.

Assume one uses, for each problem instance with m stations, the $2k+1 = 2 \cdot \sqrt{m} + 1$ diagonal strategies

$$SD_{-k}, SD_{-k+1}, \ldots, SD_0, SD_1, \ldots, SD_k$$

and gets $2k+1$ different solutions in this way. To calculate the average cost of these solutions, one first sums all their costs and then one divides the resulting sum by $2k+1$. The cost of the solution provided by SD_i or by SD_{-i} is exactly

$$m + i + \text{the number of obstacles on } D_i(D_{-i}) ,$$

for the following reasons:

(i) SD_i (SD_{-i}) uses exactly i vertical and i horizontal edges on the border in order to reach the diagonal D_i (D_{-i}) from the start and to reach the destination from the end of the diagonal. In this way, each of the tasks A_1 and A_2 is delayed exactly by i time units.

(ii) To bypass an obstacle, SD_i (SD_{-i}) uses exactly one horizontal and one vertical edge. This delays processing of A_1 and A_2 by exactly one time unit.

Let SUM denote the sum of all delays of the $2k+1$ solutions. The reader who does not like mathematical calculations can abstain from reading the next few lines.

$$\text{SUM} = \sum_{i=-k}^{k} (|i| + \text{the number of obstacles on } D_i)$$

$$= \sum_{i=-k}^{k} |i| + \sum_{i=-k}^{k} \text{the number of obstacles on } D_i$$

$$\leq 2 \cdot \sum_{i=1}^{k} i + m$$

{This was the most important step of the calculation. Since the overall number of obstacles is m, the overall number of obstacles lying on the $2k+1$ diagonals considered is at most m. Hence, the second sum is upperbounded by m.}

$$= 2 \cdot \frac{k \cdot (k+1)}{2} + m$$

{This comes from the observation of the young Gauss that the sum of the first k positive integers is exactly $k \cdot (k+1)/2$.}

$$= k \cdot (k+1) + m = k^2 + k + m = (\sqrt{m})^2 + \sqrt{m} + m$$

$$= 2m + \sqrt{m}.$$

Dividing the sum of all delays by the number of solutions, one obtains the average delay

$$\frac{2m + \sqrt{m}}{2 \cdot \sqrt{m} + 1} \leq \sqrt{m} + \frac{1}{2} = k + \frac{1}{2}\ .$$

Hence, there is at least one solution that is delayed by at most k time units[11]. Therefore, the tasks of the problem instance can be executed in at most

$$m + \sqrt{m}$$

time units.

Exercise 10.15 (Challenge) We used exactly $2 \cdot \sqrt{m} + 1$ diagonals for our calculations. What happens if one takes $4 \cdot \sqrt{m} + 1$ or $\sqrt{m} + 1$ diagonal strategies into account?

Exercise 10.16 Estimate the average delay of the 7 diagonal strategies

$$SD_{-3}, SD_{-2}, \ldots, SD_0, \ldots, SD_3$$

for the problem instance in Fig. 10.14.

We proved that each problem instance of two tasks on m stations can be solved with at most $\sqrt{m}$ delays. We learnt that online algorithms cannot avoid causing an overall delay of $m/4 = 0.25m$ time units. For large numbers m, $m/4$ can be essentially larger than $\sqrt{m}$. To achieve our result, we applied a simple combinatorial argument. In spite of its simplicity, this argument is a very powerful and successful tool of mathematics. This combinatorial method can be formulated as follows:

> *If one has m objects, each object has an assigned value, and d is the average of these values, there then exists an object with a value smaller than or equal to d, and there exists an object with a value of at least d.*

To foil the adversary (who was the winner in this game until now), we apply randomization and design a successful randomized online algorithm for this problem. First we observe that all SD_i's are online algorithms. Each diagonal strategy follows its diagonal and bypasses the obstacles in the same way, independent of their distribution. For decisions based on diagonal strategies, only knowledge about the upcoming jobs is necessary. Why is this observation helpful? We take a randomized algorithm D that

[11] If all solutions are delayed by at least $k + 1$ time units, then $k + 1/2$ cannot be the average delay.

for any problem instance with m stations, chooses one of the $2\sqrt{m}+1$ diagonal strategies SD_i at random and then applies only this strategy for calculating the solution.

Since the average delay of all $s\sqrt{m}+1$ diagonal strategies is $\sqrt{m}+1/2$, one can expect a good solution from D. Certainly, it can happen that a bad strategy for a given input instance is chosen at random. For instance, if the online algorithm D chooses the diagonal strategy SD_0 for the problem instance

$$A_1 = (1, 2, 3, \ldots, m) = A_2 = (1, 2, 3, \ldots, m)$$

at random, all m obstacles lie on the main diagonal D_0. The resulting overall delay is the worst possible, namely m. But this bad choice happens only with probability

$$\frac{1}{2\sqrt{m}+1}.$$

Since all other diagonals do not contain any obstacle, the choice of any diagonal different from D_0 may be viewed as convenient. For each i, the overall delay of SD_i is exactly i in this case.

Exercise 10.17 Let $m = 8$. Find a problem instance such that all squares of the diagonal D_3 contain an obstacle. Estimate the delays of each one of the diagonal strategies for this problem instance.

How to convince somebody using a mathematical argument that the randomized online algorithm D is practical? Is it not possible that the resulting solutions are frequently (with high probability) bad? To argue for the high quality of the designed randomized algorithm D, we apply an elegant combinatorial idea.

Consider n objects, each has assigned a positive value. In our case, the objects are particular solutions of the diagonal strategies and the values are their delays. The claim used is:

At least half of the objects have a value at most twice the average.

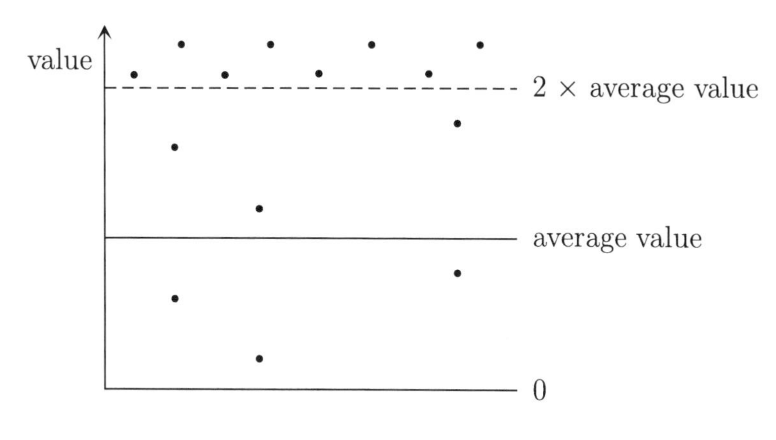

Fig. 10.15

How to argue for that? Let d be the average. If more than half of the objects have a value greater than $2d$, then d cannot be the average even if all other values are 0 (Fig. 10.15).

One can prove it also using a calculation. Let g be the number of objects with values larger than $2d$, and let h be the number of objects with values at most $2d$. Certainly, $g + h$ is the number of objects. Hence, the sum of all values of the $g+h$ objects is greater than

$$g \cdot 2d + h \cdot 0 = 2dg \ .$$

Consequently, the average d is at least

$$\frac{2dg}{g+h} \ .$$

From $d \geq 2dg/(g+h)$, one obtains

$$\begin{aligned} d \cdot (g+h) &\geq 2dg \mid \cdot \frac{1}{d} \\ g+h &\geq 2g \mid -g \\ h &\geq g \ . \end{aligned}$$

We conclude that the number g of objects with values larger than $2d$ (twice the average) cannot be greater than the number h of objects with values at most twice the average. The fact $h \geq g$ ensures that the probability of randomly choosing a strategy providing a solution with at most

$$2 \cdot d = 2 \cdot (\sqrt{m} + 1/2)$$

delays is at least 1/2. If one is not satisfied with this guarantee, one can apply the following theorem:

> *The number of objects with values larger than c times the average is at most the c-th portion of the objects.*

If $c = 2$, one obtains the claim we presented above. The number c can be chosen to be any real number greater than 1.

Exercise 10.18 (Challenge) Argue for the validity of the generalized combinatorial claim.

Applying this combinatorial claim for $c = 4$, one ensures that the probability of finding a solution causing at most $4d$ delays is at least 3/4.

Exercise 10.19 What guarantee on the overall delay of a solution computed by D can one obtain with probability at least 9/10? What is the upper bound on the number of delays for the 7/8-th solutions with the smallest number of delays?

10.4 Summary or How to Foil an Adversary

Many online problems have to be solved every day. This is the case especially for service centers that cannot foresee future requests, not even their amount and structure. In spite of that they have to take decisions with respect to known requests without knowing the next request that may come sooner or later. In many situations, it is difficult to take a decision, because current decisions can unfavorably influence how we process later requests.

The task of an algorithm designer is to recognize for which kinds of online problems one can take decisions that are as good using online strategies as one can take by knowing the future (all future requests). To analyze different scenarios, one considers the game between an algorithm designer and his or her adversary. The designer tries to design online algorithms and the adversary searches for problem instances for which the algorithms fail to offer reasonable solutions. The rule is that the adversary knows the proposed algorithm A and tries to design the future in such a way that A fails. Hence, the adversary is very powerful in this game. Consequently, it is not easy to find successful online strategies.

In this framework, randomization can be very helpful. One can explain it from the game point of view as follows. If one has a randomized online algorithm, then the main advantage of the adversary is gone. Though the adversary knows the designed online strategy, he cannot calculate the actions of the algorithm in advance, because the decisions of the randomized algorithm are taken at random during the operation of the algorithm. In our scheduling example the adversary is forced to play against $2 \cdot \sqrt{m} + 1$ strategies at once, because he does not have any chance to guess which one will be applied. As we have learned, it is even impossible to construct a problem instance that would be hard for most of the strategies. This reminds us of a football match, in which one team A follows a fixed strategy, no matter what happens during the match. If the coach of another team B sees through this strategy, he or she has a good chance to lead his or her team to victory. But if the first team A is very flexible, improvises well, and frequently performs unexpected actions, then the coach of team B searching for a winning strategy has a very hard job.

From the point of view of randomization, one can be successful by choosing one of a collection of deterministic strategies at random, if each of the strategies in this collection behaves well for most problem instances and fails on only a few of them. If each strategy fails for another small group of problem instances, then a random choice of a strategy for a given problem instance can ensure the calculation of a high-quality solution with high probability.

Solutions to Some Exercises

Exercise 10.2 For each input instance I,

$$\text{cost}(\text{Sol}_A(I)) - \text{Opt}_U(I)$$

is the absolute difference between the optimal cost and the cost of the solution computed using the algorithm A. Then the value

$$\frac{\text{cost}(\text{Sol}_A(I)) - \text{Opt}_U(I)}{\text{Opt}_U(I)} \cdot 100$$

gives the deviation of the cost of the computed solution from the optimal cost in percent.

Exercise 10.4 One has pages $1, 3, 5$, and 7 in cache and, if forced, one removes the page with the smallest number. For this strategy, the adversary constructs the following problem instance:

$$2, 1, 2, 1, 2, 1, 2, 1, 2, 1 \ .$$

The online strategy of removing the page with the smallest number results in the following solution:

$$\begin{array}{ccccccccc} 1 \leftrightarrow 2 & , & 2 \leftrightarrow 1 & , & 1 \leftrightarrow 2 & , & 2 \leftrightarrow 1 & , & 1 \leftrightarrow 2 \\ 2 \leftrightarrow 1 & , & 1 \leftrightarrow 2 & , & 2 \leftrightarrow 1 & , & 1 \leftrightarrow 2 & , & 2 \leftrightarrow 1 \end{array}$$

Clearly,

$$5 \leftrightarrow 1 \ , \ \bullet \ , \ \bullet \ , \ \bullet \ , \ \bullet \ , \ \bullet \ , \ \bullet \ , \ \bullet \ , \ \bullet \ , \ \bullet$$

is an optimal solution.

Exercise 10.6 For the problem instance $A_1 = (1, 2, 3, 4)$ and $A_2 = (3, 2, 1, 4)$, the following scheduling

time units	1	2	3	4	5
A_1	S_1	S_2	S_3	S_4	
A_2		S_3	S_2	S_1	S_4

is an optimal solution.

Exercise 10.16 All 9 obstacles are placed on the 7 diagonals considered (Fig. 10.14). D_0 contains three obstacles and so the overall delay of D_0 is $d_0 = 3$. The diagonals D_1 and D_{-1} do not contain any obstacle and their overall delay is 1 for each ($d_1 = 1, d_{-1} = 1$). This delay is caused by reaching or leaving the diagonal. The diagonals D_2 and D_{-2} contain one obstacle each. Hence, their overall delays are $d_2 = d_{-2} = 3$. Two obstacles are placed on D_3 as well as on D_{-3}, and so $d_3 = d_{-3} = 3 + 2 = 5$. The average delay on these 7 online strategies is

$$\frac{d_0 + d_1 + d_{-1} + d_2 + d_{-2} + d_3 + d_{-3}}{7} = \frac{3 + 1 + 1 + 3 + 3 + 5 + 5}{7} = 3 \ .$$

Exercise 10.17 The problem instance $A_1 = (1, 2, 3, 4, 5, 6, 7, 8, 9)$ and $A_2 = (4, 5, 6, 7, 8, 9, 1, 2, 3)$ has the property that all 6 squares of diagonal D_3 contain an obstacle.

References

[BB07] D. Bongartz and H.-J. Böckenhauer. *Algorithmic Aspects of Bioinformatics*. Springer, 2007.

[DH76] W. Diffie and M. E. Hellman. New directions in cryptography. *IEEE Trans. Information Theory*, IT-22(6):644–654, 1976.

[Die04] M. Dietzfelbinger. *Primality Testing in Polynomial Time, From Randomized Algorithms to "PRIMES is in P"*, volume 3000 of *Lecture Notes in Computer Science*. Springer-Verlag, Berlin, 2004.

[DK07] H. Delfs and H. Knebl. *Introduction to Cryptography*. Information Security and Cryptography, Principles and applications, 2nd edition. Springer-Verlag, Berlin, 2007.

[Drl92] K. Drlica. *Understanding DNA and Gene Cloning. A Guide for the Curious*. John Wiley and Sons, New York, 1992.

[Fey61] R. P. Feynman. There's plenty of room at the bottom. In: Miniaturization. *D.H. Gilbert (ed)*, pages 282–296, 1961.

[Hro97] J. Hromkovič. *Communication Complexity and Parallel Computing*. Texts in Theoretical Computer Science. An EATCS Series. Springer-Verlag, Berlin, 1997.

[Hro04a] J. Hromkovič. *Algorithmics for Hard Problems*. 2nd edition, Springer Verlag, 2004.

[Hro04b] J. Hromkovič. *Theoretical Computer Science*. Springer-Verlag, Berlin, 2004.

[Hro05] J. Hromkovič. *Design and Analysis of Randomized Algorithms. Introduction to Design Paradigms*. Texts in Theoretical Computer Science. An EATCS Series. Springer-Verlag, Berlin, 2005.

[KN97] E. Kushilevitz and N. Nisan. *Communication Complexity*. Cambridge University Press, Cambridge, 1997.

[MR95] R. Motwani and P. Raghavan. *Randomized Algorithms*. Cambridge University Press, Cambridge, 1995.

[PRS05] G. Păun, G. Rozenberg, and A. Salomaa. *DNA Computing. New Computing Paradigms*. Springer Verlag, 2005.

[RSA78] R. L. Rivest, A. Shamir, and L. Adleman. A method for obtaining digital signatures and public-key cryptosystems. *Comm. ACM*, 21(2):120–126, 1978.

[Sal96] A. Salomaa. *Public-Key Cryptography*. Texts in Theoretical Computer Science. An EATCS Series. Springer-Verlag, Berlin, second edition, 1996.

J. Hromkovič, *Algorithmic Adventures*, DOI 10.1007/978-3-540-85986-4,

Index

J. Hromkovič, *Algorithmic Adventures*, DOI 10.1007/978-3-540-85986-4,